T0135519

Oil Droplet Impact Dynamics in Aero-Engine Bearing Chambers- Correlations derived from Direct Numerical Simulations

Zur Erlangung des akademischen Grades eines

Doktors der Ingenieurwissenschaften

von der Fakultät für Maschinenbau des
Karlsruher Instituts für Technologie

genehmigte

Dissertation

von

Dipl.-Ing. Davide Peduto
aus Mannheim

Tag der mündlichen Prüfung:	30.04.2015
Hauptreferent:	Prof. Dr.-Ing. Hans-Jörg Bauer
Korreferent:	Prof. Dr.-Ing. Cameron Tropea
Korreferent:	Dr. David Hann

Forschungsberichte aus dem
Institut für Thermische Strömungsmaschinen

herausgegeben von:
Prof. Dr.-Ing. Hans-Jörg Bauer,
Lehrstuhl und Institut für Thermische Strömungsmaschinen
Karlsruher Institut für Technologie (KIT)
Kaiserstr. 12
D-76131 Karlsruhe

Bibliografische Information der Deutschen Nationalbibliothek

Die Deutsche Nationalbibliothek verzeichnet diese Publikation in der
Deutschen Nationalbibliografie; detaillierte bibliografische Daten sind
im Internet über http://dnb.d-nb.de abrufbar.

ISSN 1615-4983
ISBN 978-3-8325-4001-2

Logos Verlag Berlin GmbH
Comeniushof, Gubener Str. 47,
10243 Berlin
Tel.: +49 030 42 85 10 90
Fax: +49 030 42 85 10 92
INTERNET: http://www.logos-verlag.de

Oil Droplet Impact Dynamics in Aero-Engine Bearing Chambers- Correlations derived from Direct Numerical Simulations

Thesis submitted to

The University of Nottingham

for the degree of
Doctor of Philosophy

Dipl.-Ing. Davide Peduto

May 2015

Vorwort des Herausgebers

Der schnelle technische Fortschritt im Turbomaschinenbau, der durch extreme technische Forderungen und starken internationalen Wettbewerb geprägt ist, verlangt einen effizienten Austausch und die Diskussion von Fachwissen und Erfahrung zwischen Universitäten und industriellen Partnern. Mit der vorliegenden Reihe haben wir versucht, ein Forum zu schaffen, das neben unseren Publikationen in Fachzeitschriften die aktuellen Forschungsergebnisse des Instituts für Thermische Strömungsmaschinen am Karlsruher Institut für Technologie (KIT) einem möglichst großen Kreis von Fachkollegen aus der Wissenschaft und vor allem auch der Praxis zugänglich macht und den Wissenstransfer intensiviert und beschleunigt.

Flugtriebwerke, stationäre Gasturbinen, Turbolader und Verdichter sind im Verbund mit den zugehörigen Anlagen faszinierende Anwendungsbereiche. Es ist nur natürlich, dass die methodischen Lösungsansätze, die neuen Messtechniken, die Laboranlagen auch zur Lösung von Problemstellungen in anderen Gebieten - hier denke ich an Otto- und Dieselmotoren, elektrische Antriebe und zahlreiche weitere Anwendungen - genutzt werden. Die effiziente, umweltfreundliche und zuverlässige Umsetzung von Energie führt zu Fragen der ein- und mehrphasigen Strömung, der Verbrennung und der Schadstoffbildung, des Wärmeübergangs sowie des Verhaltens metallischer und keramischer Materialien und Verbundwerkstoffe. Sie stehen im Mittelpunkt ausgedehnter theoretischer und experimenteller Arbeiten, die im Rahmen nationaler und internationaler Forschungsprogramme in Kooperation mit Partnern aus Industrie, Universitäten und anderen Forschungseinrichtungen durchgeführt werden.

Es sollte nicht unerwähnt bleiben, dass alle Arbeiten durch enge Kooperation innerhalb des Instituts geprägt sind. Nicht ohne Grund ist der Beitrag der Werkstätten, der Technik-, der Rechner- und Verwaltungsabteilungen besonders hervorzuheben. Diplomanden und Hilfsassistenten tragen mit ihren Ideen Wesentliches bei, und natürlich ist es der stets freundschaftlich fordernde wissenschaftliche Austausch zwischen den Forschergruppen des Instituts, der zur gleichbleibend hohen Qualität der Arbeiten entscheidend beiträgt. Dabei sind wir für die Unterstützung unserer Förderer außerordentlich dankbar.

Aufbauend auf den Ergebnissen früherer Arbeiten befasst sich der Autor des vorliegenden Bandes der Schriftenreihe mit dem Aufprall von Einzeltropfen auf Flüssigkeitsfilme unterschiedlicher Dicke. Dieser Prozess besitzt hohe Relevanz für die in Lagerkammern von Flugtriebwerken auftretende Zweiphasenströmungen und dort insbesondere auf den Wärmeübergang zwischen dem schubspannungsgetriebenen Wandfilm und der Bauteilwand. Mit Hilfe des kombinierten Einsatzes moderner numerischer Verfahren und ausgewählter experimenteller Untersuchungen zeigt der Autor erstmals den Einfluss der Froude-Zahl, der dimensionslosen Filmdicke und des Aufprallwinkels auf den Eindringvorgang des Tropfens und die gebildeten Sekundärtropfen auf. Die direkte numerische Simulation des Aufprallvorgangs basiert auf der Methode der „Volume of Fluid"(VoF). Zur Begrenzung des numerischen Aufwands wird das Rechengitter gradientenbasiert an der Grenzfläche adaptiv verfeinert. Für die Ausbildung der Kalotte und der Vorhersage der Eindringtiefe genügt der Einsatz einer zweidimensionalen Rechnung, während die Entstehung einer Krone und die Bildung von Sekundärtropfen dreidimensional berechnet werden.

Auf Basis einer Vielzahl von Ergebnissen leitet der Autor Korrelationen ab, die sich für die Berechnung der gesamten Zweiphasenströmung in Lagerkammern mit Hilfe konventioneller CFD Verfahren eignen und die dabei den relevanten Parameterraum der Einflussgrößen vollständig abdecken.

Karlsruhe, im Mai 2015   Hans-Jörg Bauer

Vorwort des Autors

Die vorliegende Arbeit entstand während meiner Tätigkeit am Institut für Thermische Strömungsmaschinen (ITS) des Karlsruher Instituts für Technologie (KIT) und am Technologiezentrum für Gasturbinenantriebskomponenten (UTC) der Universität Nottingham in Großbritannien. Die Inhalte dieser Arbeit wurden im Rahmen einer Kooperationspromotion an diesen Instituten bearbeitet.

Während meiner Tätigkeit am KIT stellte mir das ITS die besten Bedingungen und eine ideale Plattform zur Anfertigung dieser Arbeit zur Verfügung. Für das Ermöglichen dieser Kooperation, für die langjährige wissenschaftliche Betreuung und für die Übernahme des Hauptreferats danke ich ganz besonders Herrn Prof. Dr.-Ing. Hans-Jörg Bauer, dem Leiter des ITS. Herrn Prof. Dr.-Ing. Cameron Tropea und Herrn Dr. habil. Ilia Roisman des Fachgebiets Strömungslehre und Aerodynamik der Technischen Universität Darmstadt gilt mein Dank für deren besonderes Interesse an meiner Arbeit und die Übernahme des Korreferats. Bei Herrn Dr. David Hann, der diese Arbeit kritisch als interner Prüfer der Universität Nottingham begutachtete, bedanke ich mich für die konstruktiven Anregungen. Die Tätigkeit am UTC in Nottingham wäre ohne die Zusammenarbeit mit Herrn Prof. Herve Morvan nicht möglich gewesen. Deshalb danke ich auch ihm ganz ausdrücklich.

Von der übergreifenden Betreuung und Unterstützung von Herrn Dr.-Ing. Rainer Koch, dem Leiter der Abteilung für Brennkammerentwicklung und Zweiphasenströmung am ITS, insbesondere von seinem Wissen und seiner wissenschaftlichen Intuition habe ich sehr profitiert. Für dieses entgegengebrachte Vertrauen danke ich diesen Kollegen und Freund sehr. Danken möchte ich auch Herrn Dr.-Ing. Klaus Dullenkopf, dem ehemaligen Leiter der Abteilung Sekundärluft -und Ölsysteme am ITS, der mich bei der Antragsstellung und Bereitstellung der finanziellen Mittel für das BMWi Vorhaben sehr unterstützte.

Meine Kolleginnen und Kollegen am ITS und am UTC haben mir stets eine sehr angenehme Arbeitsumgebung bereitgestellt, die mir jeden Tag aufs Neue Motivation und Ansporn für meine wissenschaftliche Tätigkeit gegeben hat. Von allen habe ich sehr viel lernen dürfen. Daher ist es an dieser Stelle unmöglich vollumfänglich einen Dank auszusprechen. Für die lukrativen Diskussionen gilt folgenden Herren ein besonderer Dank: Dipl.-Ing. Sebastian Gepperth, Dipl.-Ing. Siniša Kontin, Dr.-Ing. Amir Hashmi und Dipl.-Ing. Matthias Krug. Auf keinen Fall sollten die vielen Diplomanden, Studienarbeiter und Hilfswissenschaftler in Konstruktion, Messtechnik, Messauswertung und Simulation unerwähnt bleiben.

Die Entwicklung verschiedener experimenteller Einrichtungen wäre ohne die bedeutende Unterstützung der mechanischen und elektrischen Werkstatt, insbesondere von Herrn Günther Jettke, Herrn Wolfgang Schimmeyer, Herrn Lothar Kowatsch und Herrn Uwe Fränkle nicht möglich gewesen.

Der Großteil des physikalischen Verständnisses in dieser Arbeit wurde mittels aufwendiger Simulationen entwickelt, die natürlich ohne eine perfekte Infrastruktur und Betreuung seitens der IT nur sehr schwer zu bewältigen wären. Vielen Dank daher an die IT Beauftragten: Herrn

Michael Lahm und Herrn Michael Lienhart. Der starke bürokratische und administrative Aufwand, der eine Kooperationspromotion mit sich trägt, ist ohne die Hilfe von exzellenten Mitarbeiterinnen in der Verwaltung ebenfalls sehr schwer. Für deren Unterstützung danke ich ganz besonders Frau Brigitte Humbert, Frau Rosa D'Aiuto und Frau Petra Geyer.

Ohne meine zwei Schwestern, Sabrina und Rosalba Peduto, meiner Mutter - Teresa d'Atri - und meinem Vater - Lorenzo Peduto, wäre dieses wichtige Lebensziel unerreichbar geblieben. Die Geduld und Liebe, die mir meine (zukünftige) Ehefrau Olga Hamann bei der Anfertigung dieser Arbeit in den vielen „ungelebten" Wochen zugebracht hat, war auch etwas ganz besonderes. Ein großes Dankeschön dafür an meine Familie.

Karlsruhe, im Mai 2015   Davide Peduto

Contents

List of Figures   iv

List of Tables   ix

List of Symbols   x

1 Introduction   1

2 State of Scientific Knowledge   5

   2.1   Parameter of Influence describing the Drop Impact Dynamics   5

   2.2   Non-Dimensional Quantities describing the Drop Impact Dynamics   6

   2.3   Drop-to-Film Interaction in Aero-Engine Bearing Chambers   7

   2.4   Phenomenological Description of the Impingement Regimes   10

   2.5   Transitional Regime Thresholds between Spreading and Splashing   15

   2.6   Secondary Droplet Generation within the Splashing Regime   21

       2.6.1   Crown and Cavity Evolution in the Splashing Regime   21

       2.6.2   Theory of Instability Driving the Break up of the Crown's Rim   26

       2.6.3   Secondary Drop Characteristics during the Drop Impact   30

   2.7   Numerical Simulation of the Single Drop Impact onto Liquid Films   35

   2.8   Modelling Spray Impact   39

   2.9   Concretization of Objectives and Aim of Investigation   42

3 Numerical Method and CFD Modelling of Single Drop Impacts   45

   3.1   Mathematical Method - Volume-Of-Fluid Method   45

   3.2   Numerical Solution Procedure   47

       3.2.1   Discretization Procedure   47

           3.2.1.1   Discretization of the Spatial Derivatives   47

           3.2.1.2   Discretization of the Temporal Derivatives   49

       3.2.2   Pressure-Velocity Coupling   51

       3.2.3   Adaptive Mesh Refinement   51

   3.3   Computational Domain and Initial/Boundary Conditions   52

       3.3.1   2D-VOF-AMR - Numerical Model for Crown and Cavity Expansion . .   53

    3.3.2   3D-VOF-AMR - Numerical Model for Splashing 55

  3.4   Concluding Discussion 58

4 Experimental Validation of the VOF-AMR Method   **59**

  4.1   Experimental Method 59

  4.2   Results and Discussion 71

    4.2.1   Study of the Transitional Threshold between Spread and Splash 71

    4.2.2   Assessment of the 3D-VOF-AMR Method 72

    4.2.3   Assessment of the Axisymmetric 2D-VOF-AMR Method 84

  4.3   Concluding Discussion 88

5 Products of Splashing Drop Impingements without Wall Effects   **89**

  5.1   Fundamental Description of a Splashing Impingement 89

  5.2   Analysis of the Instability driving the Disintegration of the Crown 92

  5.3   Evaluation Procedure for the Following Parametric Study 96

  5.4   Effect of the Impinging Weber number 97

  5.5   Effect of the Froude number Fr 102

  5.6   Effect of the Impinging Ohnesorge Number 108

  5.7   Effect of the Impingement Angle α 113

  5.8   Concluding Discussion 119

6 Products of Splashing Drop Impingements with Wall Effects   **121**

  6.1   Cavity Penetration during Deep Pool Drop Impact 121

    6.1.1   Fundamental Description of the Cavity Evolution 121

    6.1.2   Effect of the Froude Number on the Cavity Penetration 126

    6.1.3   Effect of the Impinging Weber number on the Cavity Penetration 129

    6.1.4   Effect of the Ohnesorge Number on the Cavity Penetration 131

    6.1.5   Effect of Impingement Angle on the Cavity Penetration 133

  6.2   Effect of Cavity Penetration & Liquid Film Height on the Secondary Drop Generation 136

    6.2.1   Relation between Cavity Penetration and Lamella Expansion 136

    6.2.2   Effect of the Ratio $H^*/\Delta_{max,deep}$ on Secondary Drop Generation 140

7 Correlations for the Products of Splashing **145**

   7.1   Requirements for the Drop-Film Interaction Model 145

   7.2   Correlations for the Maximum Cavity Penetration and the Lamella Rim Height   147

   7.3   Correlating the Secondary Drop Characteristics and Ejected Mass Fraction . . . 151

   7.4   Drop-Film Interaction Model 155

   7.5   Significance of the Model for Numerical Simulation of Aero-Engine Oil Systems 156

8 Summary and Outlook **157**

Literature **161**

Appendix **175**

   A.1   First Part 175

   A.2   Second Part 179

   A.3   Spray-Film Interaction Model derived from Single Drop Impingement Data . . 179

List of Figures

1.1 The internal air and oil system in aero-engines 2

2.1 The flow phenomena in aero-engine bearing chambers (see Gorse (2007)) . . . 8

2.2 Range of impingement conditions identified in the bearing chambers 9

2.3 Depiction of Dense and Dilute Spray Characteristics, (a) from an Air Atomizing Nozzle, (b) from a Pulsed Pressurized Nozzle (images from Kontin (2015)) . . 11

2.4 Impact regimes during the single drop impingement onto liquid surfaces 12

2.5 Drop impact in the splashing regime: Water, D=2.8 mm, V=4.2 m/s, We=688, Fr=619, Re=14624 (images from Bisighini (2010)) 13

2.6 Drop impact onto wetted walls ($H^* \cong 0.1$): Comparison of available splashing thresholds with bearing chamber impact conditions 16

2.7 Application of various non splash-splash correlations to the parameter range in the bearing chamber 20

2.8 Crown dynamics during water drop impact onto a liquid surface (Image from present PhD thesis) 22

2.9 Visualization of the cavity for a acetic acid drop impact onto a deep liquid pool (Shadowgraph from Bisighini (2010)): We = 2177, Fr = 691, Re = 12642 23

2.10 Validity of the theoretical formulation of Bisighini et al. (2010) for the maximum cavity depth through comparison with experimental results in the literature and the parameter range in the bearing chambers 25

2.11 Crown disintegration during the impact of a silicone oil droplet (Re=966, We=874, $H^* = 0.2$). The image is taken from the publication of Zhang et al. (2010). . . . 26

2.12 Crown splash during a single drop impact (Re=894, We=722, $H^* = 0.2$). Relation between the Rayleigh-Plateau instability wavelength and the number of secondary droplets. Images are taken from the publication of Zhang et al. (2010) at $t = 1.85\ ms$ and $t = 3.15\ ms$ after impact. 27

2.13 Secondary droplet produced from the prompt and the corona splashing regime during single drop impingement onto a liquid layer of $H^* = 0.2$. Diagram from Zhang et al. (2010). No splash is represented by the black small circles, crown droplets with and without microdroplets by the open circles, and microdroplets without crown droplets by the diamonds. The filled squares indicate the parameter set for all experiments at $H^* = 0.2$ conducted in Zhang et al. (2010). 30

2.14 Drop-wall and drop-film interaction model comparison via the ejected mass fraction for a typical bearing chamber oil drop impact scenario 41

3.1   Boundary Conditions and Computational Mesh for the investigation of the cavity
      and lamella rim evolution   53

3.2   Computational domain, boundary conditions and a typical 3D mesh for the normal
      and oblique impingement simulations   56

4.1   Experimental setup for droplet impingement experimentation   60

4.2   Dependence of the depth-of-field on the numeric aperture and the circle of confusion 63

4.3   Typical grey-scale intensity values (b) over the central row of a falling drop (a) .   65

4.4   Principle process of grey-scale image enhancement using a background correction 66

4.5   Principle of the post processing routine   67

4.6   La Vision calibration plate used for the depth-of-field calibration   69

4.7   Non-dimensional sizing accuracy for different out-of-focus positions   70

4.8   Calibrated normalized intensity gradients as a function of z and the reference
      diameter   70

4.9   Various splashing limits for single drop impact onto shallow and deep liquid pool
      ($H^* > 1$)   72

4.10 Drop impact onto a deep liquid pool: Role of time during the impact evolution
      (We=800, Oh=0.0018)   74

4.11 Comparison of a 3D-VOF-AMR simulation (left: iso-surface at $\alpha = 0.5$) with
      shadowgraphs (right) at different impingement stages: water, $D = 4.2\ mm$,
      $V = 3.7\ m/s$ (case 3)   75

4.12 Motion of secondary drops in the impingement region during a splashing water
      drop impingement onto a deep liquid pool: $D = 4.2\ mm$ and $V = 3.7\ m/s$ . .   76

4.13 Typical relation between the secondary drop mass and secondary drop diameter
      for a water drop impingement of $D = 4.2\ mm$ and $V = 3.7\ m/s$ 77

4.14 Fitting of a log-normal distribution function onto the measured secondary drop
      diameter distribution of a splashing water drop impingement onto a deep liquid
      pool: $D = 4.2\ mm$ and $V = 3.7\ m/s$ 78

4.15 Measured arithmetic mean diameter and velocity of secondary drops versus
      the K number of all the investigated test cases and comparison with available
      correlations in the scientific literature   79

4.16 Comparison of the calculated number distribution of the secondary drop sizes
      during single drop impingement onto a deep pool with the experimental results   80

4.17 Comparison of the calculated size and velocity distributions of the secondary
      droplets during single drop impingement onto a deep pool with experimental
      results and the correlation of Okawa et al. (2006)   81

4.18 Effect of the We, the Oh and the K number on secondary drop mass and number during single drop impingement on deep liquid pools and comparison with available correlations in the literature 82

4.19 Comparison of the ejected secondary drop mass and number during single drop impingement onto a deep pool with the experimental results of chapter 4 and the correlation of Okawa et al. (2006) 83

4.20 Comparison of a 2D axisymmetric VOF simulation (left: iso-line at $\alpha = 0.5$) with shadowgraphs (right) of Bisighini (2010) at different impingement stages (see case 6 in Table 4.3 for details) 85

4.21 Comparison of cavity and lamella rim evolution with experimental results of Liow (2001), Bisighini (2010) and Chapter 4 (see details in Table 4.3) 86

5.1 Fig. is continued in the next page 90

5.2 Iso-surface of $\alpha = 0.5$ showing different stages of the crown's breakup during drop impact on a liquid layer (We=2209, Oh=0.0031) 91

5.3 Different parameters describing the crown's rim 92

5.4 Iso-surface of $\alpha = 0.5$ coloured by the static pressure and absolute velocity during a crown's breakup: We=2209, Oh=0.0031, and Fr=343 93

5.5 Capillary corrugations perturbing the crown's rim (We=2209, Oh=0.0031 and Fr=343) 93

5.6 Typical parameter describing the crown's instability during single drop impact onto liquid surfaces (We=2209, Oh=0.0031 and Fr=343) 95

5.7 Qualitative depiction of the Weber number influence on the crown's breakup process during single drop impingement onto liquid surfaces 98

5.8 Effect of Weber number on the evolution of the lamella rim: $Oh = 0.0161$, $D = 0.175\,mm$ 99

5.9 Effect of the Weber number on the characteristics and mass of the secondary drops during single drop impact onto deep pools 100

5.10 Qualitative depiction of the influence of the Froude number on the crown's breakup process during single drop impingement onto liquid surfaces 104

5.11 Effect of Froude number on the evolution of the lamella rim: $We = 1787$, $Oh = 0.0161, D = 0.175\,mm, V = 15\,m/s$ 105

5.12 Effect of Froude number on characteristics and mass of secondary drops during single drop impact onto liquid surfaces 106

5.13 Qualitative depiction of the influence of the Ohnesorge number on the crown's breakup process during single drop impingement onto liquid surfaces: $We \approx 2300$ and $H^* >> 6$ 109

5.14 Effect of Ohnesorge number on the evolution of the lamella rim during single drop impingement onto a deep liquid pool: $We = 1787$, $Fr = 131061$, $D = 0.175$ mm, $V = 15\,m/s$ 110

5.15 Effect of Ohnesorge number on the characteristics and mass of the secondary drops during single drop impact onto liquid surfaces 112

5.16 Angle description for oblique impingement simulations 113

5.17 Qualitative depiction of the influence of the impact angle on the crown's breakup process during single drop impingement onto liquid surfaces from the side (We=1850, Oh=0.036, Fr=195072) 115

5.18 Qualitative depiction of the influence of the impact angle on the crown's breakup process during single drop impingement onto liquid surfaces from the top (We=1850, Oh=0.036, Fr=195072) 116

5.19 Effect of impingement angle on the evolution of the lamella rim: $We = 542$, $Fr = 131061$, $Oh = 0.0088$ 117

5.20 Effect of impact angle on the characteristics and mass of the secondary drops during single drop impact onto liquid surfaces 118

6.1 Iso-surface of $\alpha = 0.5$ showing the temporal evolution of the cavity and lamella rim during drop impact on a liquid layer: We=541, Oh=0.0161, D=0.175 mm and V=15 m/s 122

6.2 Typical cavity and lamella rim expansion over time during a submillimeter drop impingement onto a deep liquid pool: $We = 1787$, $Oh = 0.0161$ and $Fr = 131061$ 123

6.3 Fig. is continued on the next page 124

6.4 Iso-surface of $\alpha = 0.5$ showing pressure and velocity field within cavity and lamella rim during drop impact on a liquid layer: We=541, Oh=0.0161, D=0.175 mm and V=15 m/s 125

6.5 Iso-surface of $\alpha = 0.5$ showing effect of the Froude number on the temporal evolution of the cavity and lamella rim during drop impact on a liquid layer: $We = 1787$, $Oh = 0.0161$, $D = 0.175\,mm$, $V = 15\,m/s$ 127

6.6 Effect of Froude number on the evolution of the cavity during single drop impingement onto a deep liquid pool: $We = 1787$, $Oh = 0.0161$, $D = 0.175\,mm$, $V = 15\,m/s$ 128

6.7 Iso-surface of $\alpha = 0.5$ showing the effect of impact velocity on the temporal evolution of the cavity: $Oh = 0.0161$, $D = 0.175\,mm$ 129

6.8 Effect of Weber number on the evolution of the cavity: $Oh = 0.0161$, $D = 0.175$ mm 130

6.9 Iso-surface of $\alpha = 0.5$ showing the effect of liquid viscosity on the temporal evolution of the cavity and lamella rim during drop impact on a liquid layer: $We = 1787$, $Fr = 131061$, $D = 0.175\,mm$, $V = 15\,m/s$ 132

6.10 Effect of Ohnesorge number on the evolution of the cavity: $We = 1787$, $Fr = 131061$, $D = 0.175\ mm$, $V = 15\ m/s$ 133

6.11 Iso-surface of $\alpha = 0.5$ showing the effect of impingement angle on the temporal evolution of the cavity and lamella rim during drop impact on a liquid layer: $We = 541$, $Fr = 131061$, $Oh = 0.0088$ 134

6.12 Effect of impingement angle on the evolution of the cavity: $We = 542$, $Fr = 131061$, $Oh = 0.0088$ 135

6.13 Iso-surface of $\alpha = 0.5$ showing the effect of liquid film height on the temporal evolution of the cavity and lamella rim during drop impact on a liquid layer: $We = 1787$, $Fr = 131061$, $Oh = 0.0161$: $D = 0.175\ mm$, $V = 15\ m/s$ . . . 137

6.14 Effect of liquid film height on the evolution of the cavity and lamella rim during single drop impingement onto a deep liquid pool: $We = 1787$, $Fr = 131061$, $Oh = 0.0161$: $D = 0.175\ mm$, $V = 15\ m/s$ 138

6.15 Qualitative depiction of the liquid film height influence on the crown's breakup process during single drop impingement onto liquid surfaces 141

6.16 Effect of liquid film height on characteristics and mass of secondary drops during single drop impact onto liquid surfaces 143

7.1 Definition of pre -and post impingement parameters needed in a drop-film interaction model 145

7.2 Sketch showing the pre -and post impingement parameter 146

7.3 Maximum crater depth $\Delta_{max,deep}$ at $\tau_{cav,max,deep}$ in deep pool impact: Comparison of all simulation results with the theoretical model of Bisighini et al. (2010) and equation 7.1 148

7.4 Maximum lamella rim height H_{cr}/D at $\tau_{cr,max}$ in single drop impingement onto liquid surfaces $H^* > 1$: Comparison of all simulation and experimental results with the predictions of equation 7.1 and 7.3 150

7.5 Graphical verification of the correlations for secondary drop mass and number during single drop impingement onto liquid surfaces 152

7.6 Graphical verification of the correlations for secondary drop sizes and velocities during single drop impingement on liquid surfaces 154

7.7 Comparison of the investigated experimental and numerical range of single drop impingement events with the impingement conditions derived from measurements in the bearing chamber 155

List of Tables

3.1 Example of the required computational resources for two different impingement conditions 54

3.2 Example of the required computational resources for one impingement using both numerical setups 57

3.3 Boundary conditions for the normal impact (left) and the oblique impact (right) 57

4.1 Characteristic parameters of the optical setup 63

4.2 Test cases for validation of the 3D-VOF-AMR method 73

4.3 Test cases for validation of the axisymmetric 2D-VOF-AMR method 84

5.1 Test conditions for the effect of Weber number 97

5.2 Test conditions for the effect of Froude number 103

5.3 Test conditions for the effect of Ohnesorge number 108

5.4 Test conditions for the effect of impingement angle 114

6.1 Test conditions for the effect of $H^*/\Delta_{max,deep}$ on the cavity and lamella expansion 136

6.2 Test conditions for the effect of liquid film height 140

7.1 Coefficients and exponents for the maximum cavity depth $\Delta_{max,deep}$ and $\tau_{cav,max,deep}$ including the uncertainty U_x in the powers 148

7.2 Coefficients and exponents for the maximum lamella rim height $H_{cr,max}/D$ and $\tau_{cr,max}$ including the uncertainty U_x in the powers 149

7.3 Coefficients and exponents for the ejected mass M_{sec}/M and number N of secondary drops during single drop impingement onto liquid surfaces for $45° < \alpha < 90°$ including the uncertainty U_x in the powers 151

7.4 Coefficients and exponents for the ejected mass M_{sec}/M and number N of secondary drops during single drop impingement onto liquid surfaces $30° < \alpha < 45°$ including the uncertainty U_x in the powers 151

7.5 Coefficients and exponents for the secondary drop sizes and velocities during single drop impingement onto liquid surfaces including the uncertainty U_x in the powers 153

A.1 Classification of the target surface for drop impingement onto liquid surfaces . 179

List of Symbols

Symbol	Description	Unit
τ	Dimensionless time	
A	Coefficient	$-$
A	Surface Area	m^2
C	Constant	$-$
c	Coefficient	$-$
D	Droplet diameter	m
d	Differential	$-$
f	Focal length	$-$
f	Force	N
f	Impact frequency	Hz
f	Probability density function	$-$
Fr	Froude number	
g	Gravitational acceleration	m/s^2
H	Height	m
h	Thickness	m
H^*	Relative film height	
K	Cossali number	
l	Length	m
M	Spray/Droplet Mass	kg
m	Magnification factor	$-$
N	Number of secondary droplets	$-$
n	Number of pixel	$-$
Oh	Ohnesorge number	
p	pressure	Pa
q	Radius of an airy disc	m
R	Radius	m
R	Wall surface characteristic	$-$
Re	Reynolds number	
S	Constant	$-$
S	Local source term	$-$
s	Pixel width	m
T	Temperature	K
t	Time	s
u	Distance object-lens	m
u,v,w	Velocity components in space	m/s
V	Drop impact velocity	m/s
v	Distance sensor-lens	m

Vol	Volume	m^3
We	Weber number	
x,y,z	Cartesian coordinates	m
Y	Cavity depth	m

Greek Symbols

α	Dimensionless crater radius	$-$
α	Impact angle	$^\circ$
α	Indicator function	$-$
Δ	Non-dimensional cavity depth	$-$
ϵ	Threshold value	$-$
Γ	Diffusion coefficient	$-$
κ	Free surface curvature	$1/m$
λ	Wave length	m
μ	Dynamic viscosity	kg/ms
∇	Gradient operator	$-$
ν	Kinematic viscosity	m^2/s
Ω	Dimensionless crater diameter	$-$
Φ	General flow quantity	$-$
ρ	Fluid density	$\frac{kg}{m^3}$
σ	Standard deviation	$-$
σ	Surface tension	N/m
τ	Shear tensor	N/m^2
ζ	Axial coordinate of the sphere center	$-$

Abreviations

AMR	Adaptive Mesh Refinement
CCD	Charged-Coupled Device
CFL	Courant-Friedrich-Lewis
$CICSAM$	Compressive Interface Capturing Scheme for Arbitrary Meshes
$CLSVOF$	Coupled Level-Set-VOF
$CMOS$	Complementary Metal Oxide Semiconductor
CoC	Circle of Confusion
CSF	Continuum Surface Force
DAS	Donor-Acceptor Scheme
DoF	Depth-of-Field
exp	Exponential function
fr	Frame rate

HS	High-Speed Shadowgraphy
MAC	Marker and Cell
NA	Numeric aperture
ODE	Ordinary differential equations
PDA	Phase Doppler Anemometer
$PISO$	Pressure Implicit with Splitting Operators
$PLIC$	Piecewise Linear Interface Capturing
PTV	Particle Image Velocimetry
px	Pixel
SPH	Smooth Particle Hydrodynamics
VOF	Volume-Of-Fluid
WG	Water Glycerine

Indices

$*$	Non-dimensional
0	Initial value
$\cdot$	First temporal derivative
σ	Indices for surface tension
$\rightarrow$	Vector operator
a	Surface roughness
c	Surface curvature
cav	Cavity
CF	Crown formation
CL	Cavity and Lamella
$Cossali$	Author: Cossali
cr	Crown
cr,max	Maximum value of crown height
$crit$	Critical
cyl	Cylinder
exp	Experiment
f	Faces
f	Liquid film
g	Gas
i,j,k	Indices in the Cartesian tensor notation
kd	Threshold Coalescence-Splashing by Samenfink et al. 1999
l	Liquid
m	Mean
max	Maximum
$max,deep$	Maximum value in deep pool impact
mod	modified

$Mundo$	Author: Mundo
n	Exponent
n	Normal
$orthogonal$	Orthogonal to undisturbed liquid surface
p,q	Indices for mean spray diameter
$parallel$	Parallel to undisturbed liquid surface
$real$	Reality
ref	Reference
SD	Secondary droplets
sec	Secondary droplets
$sensor$	Sensor
T	Transposed
th	Theory

1 Introduction

The growing need for global transportation, and its increasing impact on the environment, makes the positioning of a competitive aircraft engine on the aviation market of the future a real challenge: more efficient aero-engines with larger power outputs but lower noise levels and emissions are demanded. These trends make imperative the perfect design of not only the main gas path but especially the transmission- and the secondary air and oil systems.

The secondary air system is fed by the compressor bleeds, cools the engine components, and seals the bearing chambers, as well as carrying out some other rather minor tasks. The oil system, on the other hand, has as its main task the lubrication and cooling of the bearings and gears, which are highly loaded. Figs 1.1a and 1.1b depict a schematic of an oil system in an aero-engine and a typical section in the rear part of the engine. This shows the complexity of the different paths of the air and oil. The lubricating/cooling oil within an oil system is stored in a tank and supplied to the bearings and gears with an under-race, under-cage, or direct jet lubrication injection. Beside the cooling and lubrication of the bearings themselves, its main task is removing particles of debris from the bearings and the gears, together with cooling the surrounding bearing compartments. As soon as the latter is accomplished, the oil flows into the adjacent chamber, commonly referred to as the bearing chamber. Bearing chambers are usually operated with labyrinth seals fed with sealing air in order to prevent oil leakage from the bearing chambers, which would cause oil coking, oil fires, or contamination of the cabin air. The supplied sealing air is derived from the internal air system. Hence, a very complex air-oil mixture flow takes place in bearing chambers: it consists of a droplet laden core air flow, an oil wall film attached to the outer chamber walls, and a variety of phase interactions. This air-oil mixture is, moreover, subject to the rotating momentum coming from the shaft. The two-phase flow evolving in the rear bearing chambers is further expected to carry the heat flux imposed by conduction from the hot environment (the turbine and the combustion chamber) to cool the compartment. This can only be achieved safely and reliably when the residence time of the oil inside the bearing chamber is designed to be as short as possible (to avoid oil degradation/coking), while still long enough to ensure sufficient convective cooling. To date, this is usually controlled by introducing two exit pipes in the bearing chamber, which are usually referred to as the scavenge and vent off-take. The flow exiting via the vent off-take is driven by the pressure gradient between the bearing chamber and the environment, whereas additional displacement pumps are needed to pump the oil out through the scavenge off-take. In the optimum case, the complete oil volume exits through the scavenge and the air through the vent off-take pipe. However, the reality shows still very complex two-phase flows in both off-take pipes, which require additional components as breathers to redirect pure oil back to the tank. To date, a safe and reliable design of the oil system has only been achieved by supplying much more oil to the bearing and gears than is actually required for lubrication and cooling. This is necessary to maintain the oil and material temperature below their critical values. The reason for this shortcoming is the lack of understanding of the flow phenomena taking place inside the oil systems.

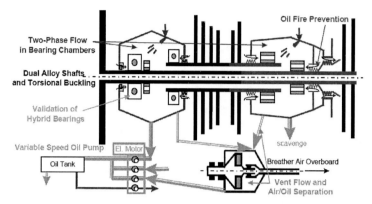

Oil Fire Prevention

Two-Phase Flow
in Bearing Chambers

Dual Alloy Shafts
and Torsional Buckling

Validation of
Hybrid Bearings

Variable Speed Oil Pump   El. Motor

Oil Tank

scavenge

Breather Air Overboard

Vent Flow and
Air/Oil Separation

(a) Schematics of a typical oil system (Klingsporn et al. (2004))

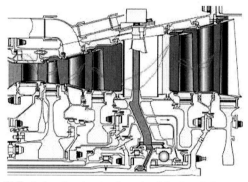

(b) Typical section showing the paths of the air and oil system
(Rolls-Royce plc)

Fig. 1.1: The internal air and oil system in aero-engines

In the last two decades, extensive research has been undertaken at the Institut für Thermische Strömungsmaschinen of the Karlsruhe Institute of Technology (KIT) and the University of Nottingham to identify the flow patterns and gain a detailed understanding of these patterns with respect to the heat transfer in model bearing chambers and generic test rigs. The most relevant parts of this research were delivered, among others, by Glahn and Wittig (1996), Glahn et al. (1997), Busam et al. (2000), Gorse et al. (2004), Chandra et al. (2010) and Kurz et al. (2012)), where all the mentioned bearing chamber flow patterns were identified and their influence on the wall heat transfer was described. The air flow, wall film, and droplet dynamics inside bearing chambers were characterized at several positions in and along the circumference of the chamber. However, general design guidelines allowing the transfer of this knowledge to real engine conditions has still not been gained, due to the strong dependence of the flow phenomenon on the geometric configuration of the bearing chambers. Moreover, a suitable instrumentation and optical access allowing a comprehensive insight into each individual flow phenomenon within the bearing chamber is nearly impossible.

Experimental setups containing real engine boundary conditions and bearing chamber geometry, on the other hand, are highly expensive and time consuming; hence, not practical. The continuous advances in CFD multiphase modelling techniques, together with improving computational resources, have enabled a new vision for future aero-engine oil system design. As an advantage compared to empirical correlations, the entire flow field may be calculated with less effort and may be studied locally in more detail. With all this upcoming potential, however, the state-of-the-art of CFD multiphase flow techniques, to date, is still not at a level to capture the detailed two-phase flow and heat transfer in the oil systems of an aero-engine, and requires still further provision. For this reason, several investigations are currently under way that aim to analyse, validate, and improve the applicability of the available multiphase modelling techniques to the prediction of the individual flow phenomena. Despite the success in modelling the droplet laden air flow with the help of the Lagrangian Particle Tracking approach, and the wall-film dynamics with the homogeneous Eulerian multiphase approach in the bearing chambers, none of the scientific studies have focussed on the mass and momentum exchange between the droplets and the liquid films. This interaction, often referred to as the wall-effect, has a decisive impact on the two-phase flow and heat transfer, and needs to be accurately integrated into future oil system CFD design tools. A strong droplet impingement rate on the wall-film disturbs the local wall-film flow at the wall, and affects the local wall heat transfer. The sprays generated by the droplets impinging on the walls and films may produce very small secondary droplets and enhance the aforementioned grave issues (oil coking, oil fire, and cabin air contamination). In the scientific literature, empirical models describing the interaction of the droplets with the walls and films are very rich, and mostly developed for applications to automotive Direct Injection Engines. As will be shown in the state-of-the-art chapter, however, their application is limited to the range of parameters for which they were derived. Most of the drop impingement models have been developed for kinematic impact conditions and fluid properties significantly differing from the operating range in bearing chambers. Moreover, particularly with respect to the mass and momentum exchange between the impinging droplets and the liquid films where information about the ejected liquid mass flow is required, most measurement techniques are limited and the results may appear to be not physically realistic (Cossali et al. (2005)).

The present investigation aims therefore at the development of a drop-film interaction model for application to the parameter range appropriate to aero-engine oil systems. With the inclusion of the present sub-model in the multiphase CFD methods in aero-engine oil system design, the wall-effects due to the drop-film interaction are successfully resolved, hence, more details of the entire bearing chamber two-phase flow and heat transfer become visible. A better numerical prediction of the bearing chamber two-phase flow together with an improved understanding of the flow structures enhances oil system design in terms of safety, reliability, and efficiency. It brings within reach, in the near future, an oil system operating with a reduced amount of oil, and smaller pipes, tanks, pumps and breathers.

2 State of Scientific Knowledge

The introduction chapter indicated the need for improved numerical simulation methods in aero-engine oil system design. One of the major demands within these methods is the prediction of the spray generation due to the drop impingement. There are mainly two reasons why this two-phase flow phenomenon has not been investigated in the bearing chamber environment. From the experimental perspective, it is nearly impossible to resolve the droplets, which interact with the liquid surfaces, due to the many coincident flow phenomena at the same time. Within the numerical methods, on the other hand, capability to resolve the flow structures may exist, but with computational efforts that are up to date still impractical. A numerical methodology, which enables the prediction of the spray generation and propagation in bearing chambers, can only be achieved by integrating semi-empirical correlations (sub-models). A set of correlations for the mass and momentum exchange between the impinging droplets and the liquid wall films is aimed at in the present thesis.

After the description of the parameter of influence and the non-dimensional quantities, which govern the drop-film interaction, the range of impingement conditions, expected in bearing chambers, will be presented with the help of the literature. Then, the drop impact dynamics and the available drop-film interaction models are described and discussed in the light of impingement conditions found in bearing chambers.

2.1 Parameter of Influence describing the Drop Impact Dynamics

Before presenting the range of drop-to-film impingement conditions, which have been identified in the bearing chambers, it is necessary to describe first all the parameters of influence. These parameters are then combined to several non-dimensional quantities, which will be used for description, scaling and modelling of the drop impact dynamics. The parameters of influence for drop impingement onto liquid wall films are briefly presented next. A more detailed description of the parameters can be found in the appendix.

- t $[s]$ Impingement time
- g $[m/s^2]$ Gravitational or centrifugal acceleration
- ρ_l $[kg/m^3]$ Liquid density
- μ_l $[kg/ms]$ Dynamic liquid viscosity
- σ $[N/m]$ Surface tension
- D $[m]$ Impacting drop diameter
- V $[m/s]$ Impacting absolute drop velocity
- α $[°]$ Impact angle
- h_f $[m]$ Liquid wall film height

2.2   Non-Dimensional Quantities describing the Drop Impact Dynamics

The non-dimensional time describes the influence of time on the drop impingement process and allows comparison of several phenomena, e.g. the cavity & crown evolution, and the breakup process in a dimensionless form.

$$\tau = \frac{t \cdot V}{D}. \tag{2.1}$$

The Weber number represents the ratio between the inertial forces and the surface tension of an impinging drop. It is a measure of the relevance of surface tension forces during the impact. This parameter is involved in the majority of the analytical and empirical descriptions of the different flow phenomena during the drop impingement evolution and has to be considered as one of the most relevant non-dimensional numbers.

$$We = \frac{\rho_l V^2 D}{\sigma}. \tag{2.2}$$

For the drop impact dynamics, the Froude number represents the ratio between the inertial forces and the gravitational acceleration. This parameter is used to examine the effect of the gravitational forces on the impingement process. In bearing chambers, despite the gravity, the centrifugal forces acting on the droplets, due to the large rotational shaft speed, are more frequent. Hence, the gravity needs to be substituted with the centrifugation.

$$Fr = \frac{V^2}{gD}. \tag{2.3}$$

The Reynolds number describes the ratio between the inertial forces and the viscous forces. It is a substantial measure of the influence of the viscous forces. By taking the ratio of the Weber number and the Reynolds number, a parameter describing the relation between the viscous forces and the surface tension is obtained. Large Ohnesorge numbers denote a larger effect of the liquid viscosity on the outcome of the drop impact. The Laplace number, $La = Oh^{-2}$, is occasionally used in literature instead of the Ohnesorge number.

$$Re = \frac{\rho_l D V}{\mu_l}, \tag{2.4}$$

$$Oh = \sqrt{\frac{We}{Re^2}} = \frac{\mu_l}{\sqrt{\rho_l \sigma D}} = \frac{1}{\sqrt{La}}. \tag{2.5}$$

An important scaling quantity, usually applied for the definition of droplet impact regimes or as a model parameter, is the K parameter, as proposed by Cossali et al. (1997) (see also Cossali et al. (2004), Tropea and Marengo (1999), Okawa et al. (2006), and Okawa et al. (2008)).

$$K_{Cossali} = We \cdot Oh^{-0.4} \qquad (2.6)$$

In principle, it is a measure for a non-dimensional impacting drop momentum.
The influence of the liquid film height or of the bottom wall on the impingement process is described by forming the ratio between the film height and the impacting drop diameter. Hence, larger relative film heights represent a reduced influence of the bottom wall on the impact evolution.

$$H^* = \frac{h_f}{D}. \qquad (2.7)$$

2.3 Drop-to-Film Interaction in Aero-Engine Bearing Chambers

In order to draw a parameter range of interest for the modelling of drop-film interaction, this section gives a very brief summary of the most known publications, which describe the internal two-phase flow in the bearing chambers.

A world unique test environment, the High Speed Bearing Chamber Test Rig, have been developed at the Institute for Thermal Turbomachinery aiming at the characterization of the bearing chamber two-phase flow, under conditions close to real aero-engines (see Wittig et al. (1994)). This test facility is capable of simulating the influence of the sealing air flow, the oil feed flow, the shaft speed, and the chamber pressure on the two-phase flow and heat transfer. A detailed description of the rig design can be found in Wittig et al. (1994), Glahn (1995), Glahn and Wittig (1996), and Gorse et al. (2004). Other useful test facilities aiming at more the exploration of the general macroscopic bearing/bearing chamber system behaviour rather than the local flow characterization are found in Flouros (2005), Flouros (2006), Simmons et al. (2011), Chandra et al. (2010), and Gloeckner (2011). Fig. 2.1 shows a summary of all the co-occurring two-phase flow phenomena in an aero-engine bearing chamber.

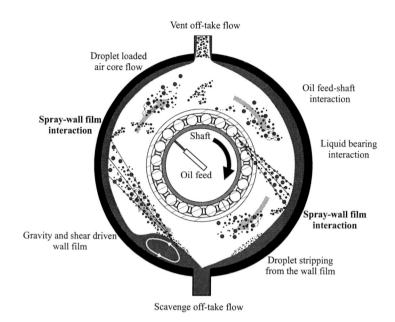

Fig. 2.1: The flow phenomena in aero-engine bearing chambers (see Gorse (2007))

In general, the research with the mentioned test facility and other more generic test rigs focussed on the characterization of

- the pure air flow dynamics in the bearing chambers (see, e.g. Gorse et al. (2003), Lee et al. (2005), Aidarinis et al. (2010), and Hohenreuther (2011)),

- the droplet dynamics in the bearing chambers (see, e.g. Birkenkaemper (1996), Glahn et al. (1997), Wittig and Busam (1996), Glahn et al. (2002), Glahn et al. (2003), Gorse et al. (2008), and Krug et al. (2015)),

- and the wall film dynamics (Wurz (1971), Sattelmeyer (1985), Himmelsbach (1992), Himmelsbach et al. (1994), Elsaesser (1998), Rosskamp (1998), Ebner (2004), Ebner et al. (2004), Schober (2009), Wittig et al. (1994), Glahn and Wittig (1996), Busam et al. (2000), Gorse et al. (2004), Kurz et al. (2012), Young and Chew (2005), Robinson and Eastwick (2010), Peduto et al. (2011a), Hashmi (2012), and Mathews et al. (2013)).

From these investigations the following conclusions and parameter range of drop-to-film interaction in aero-engine bearing chambers can be derived:

- The sprays in the bearing chambers are more dilute than the diesel engine sprays; they mostly interact with liquid rather than dry surfaces. The drop sizes in the bearing chambers are in the submillimetre range (see Glahn et al. (1997), Glahn et al. (2002), Glahn et al. (2003), and Krug et al. (2015)). Fig. 2.2b confirms the latter statement by showing very high Froude numbers Fr, which can principally be related to the submillimeter drop sizes in the bearing chambers.

- In the last decades, the height of the wall-films in the bearing chambers has been intensively analysed with the aim to characterize the convective internal heat transfer. With respect to drop-film interaction, it has been shown that the heights of the wall-films are much larger than the drop sizes. However, drop sizes larger than $D > 500\ \mu m$ occasionally exist in the bearing chambers. For the impingement of these droplets, the wall affects the impingement evolution. Hence, the influence of the wall during the drop interaction with wall films has to be taken into account for drop-film interaction modelling in the bearing chambers. This fact is also indicated by the $K - H^*$ map in Fig. 2.2c, which shows the range of the possible relative film heights for the different impinging drop momentums K.

- The smaller drops in bearing chambers have larger velocities, follow the air flow, and impinge with smaller angle (from horizontal) onto the liquid surface. Larger drops, in contrast, are slower, leave the air flow streamlines sooner, and are carried by the centrifugal forces, which results into steeper impingement angles α. Hence, there is a significant change in the impingement angle in the bearing chambers, as a function of the droplet size, which may affect the impact outcome. Fig. 2.2d depicts the latter fact by showing the drop momentum K versus the possible impingement angles.

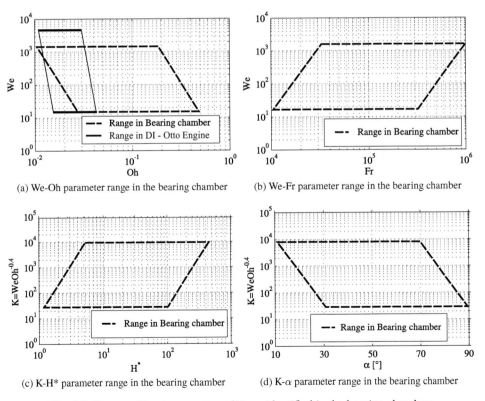

(a) We-Oh parameter range in the bearing chamber

(b) We-Fr parameter range in the bearing chamber

(c) K-H* parameter range in the bearing chamber

(d) K-α parameter range in the bearing chamber

Fig. 2.2: Range of impingement conditions identified in the bearing chambers

- The fluid properties in the bearing chambers significantly vary with the operation of the aero-engine (internal temperature). The viscosity values, for instance, show a very large range, extending over several orders of magnitude. This, in turn, leads to a very unique range of impingement conditions, dissimilar to those detected in internal combustion engines, where most of the literature for drop impact dynamics has been directed.

Based on the reviewed studies for bearing chambers, Figs 2.2a-2.2d summarize a relevant non-dimensional parameter range for drop impact dynamics in the bearing chambers. This parameter range is used in the following work. Moreover, in Fig. 2.2a, a comparison between the impinging We-Oh map, found in the Direct Injection Otto Engines and the aero-engine bearing chambers, is shown. As can be seen, in the bearing chambers, the drop impingement can take place with a much larger Ohnesorge number than with DI Otto Engines. Since most of the drop-film interaction models have originally been developed for DI Otto Engines applications, the need for the drop-film interaction models, which are applicable to the impingement conditions found in the bearing chambers, is claimed. Hence, with the aid of the identified parameter range of drop-film interaction in bearing chamber, the base to discuss the state-of-the-art of the droplet impact dynamics and the spray impingement modelling in the next section is set.

2.4 Phenomenological Description of the Impingement Regimes

In most of the practical situations, drops interact with walls and films within a spray, which is either dense or dilute. A dilute spray is characterized by a low liquid mass concentration in the air, where the droplets do not interact with each other before and after the impact. From the modelling point of view, these two impact regimes have to be treated using different approaches (see Tropea and Marengo (1999)). Fig. 2.3 exemplarily depicts the difference between a dilute and a dense spray. The droplets impacting in a dense spray (e.g. direct injection of Diesel fuel) are influenced by the temporal and spatial interaction of the surrounding impingement events (see Kalantari (2007) and Richter et al. (2005)). The modelling approaches here rely on empirical models derived from poly-disperse spray impact scenarios, which are very scarce in the literature. For the impact of droplets from a very dilute spray, on the other hand, the influence of the temporal and spatial interaction of the impinging drops plays a minor role. The spray impact models, derived from the fundamental studies of single drop/mono-dispersed drop stream impingements, are appropriate (see Roisman et al. (2006)). In the present thesis, the focus is directed to the dilute spray impingement, since sprays in the bearing chambers are rather dilute. This thesis mainly concentrates therefore on the single drop impacts and the spray impact modelling derived from the single drop impingement data. The discussion of the most relevant literature, dedicated to the single drop impact dynamics, is conducted next.

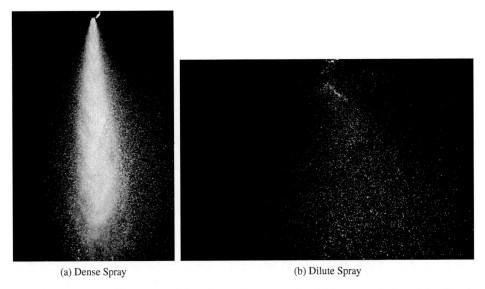

(a) Dense Spray   (b) Dilute Spray

Fig. 2.3: Depiction of Dense and Dilute Spray Characteristics, (a) from an Air Atomizing Nozzle, (b) from a Pulsed Pressurized Nozzle (images from Kontin (2015))

The impact of single drops onto walls and films is a natural phenomenon, which has attracted the attention of scientists and artists since 1508. Da Vinci (1508) was one of the first authors fascinated by the shapes produced as the result of an impact of a droplet onto a surface. About three hundred years later, Worthington (1876) developed more quantitative studies on this flow phenomenon. This sets the backdrop for a tradition of research not only of fundamental scientific interest, but also relevant to many applications in ecological, agricultural, meteorological, and engineering fields. Although a lot of research has been undertaken in this field, up to date, a complete and quantitative description of the physical processes of the drop impact dynamics is still lacking. In principle, the available literature on the single drop impact can be divided into three main impingement scenarios, specifically the impact of droplets onto (i) cold walls, (ii) hot walls, and (iii) films. Due to the large amount of the literature, available for each regime, the material to be discussed here needs to be limited to the most common impact scenarios, which occur in aero-engine bearing chambers. Hence, the emphasis is put on the single drop impact onto liquid films. For more details about single droplet impingement onto dry cold, and hot walls, please refer to the publications of, e.g. Richter et al. (2005) and Mueller et al. (2009). Fig. 2.4 qualitatively displays the different outcomes for a single drop impact onto liquid films.

When a single droplet impinges on a liquid wall film, different impingement outcomes may result. Bai and Gosman (1995) defined seven possible impact regimes, which an impinging droplet may undergo. The four most relevant regimes are the stick/deposition, the rebound, the spread, and the splashing regime. For impact onto thick liquid films, Bisighini (2010) found a more appropriate description of the impingement regimes with the coalescence, the cratering, and the splashing regime.

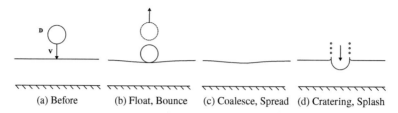

(a) Before   (b) Float, Bounce   (c) Coalesce, Spread   (d) Cratering, Splash

Fig. 2.4: Impact regimes during the single drop impingement onto liquid surfaces

The following description of the impingement regimes assumes a Ohnesorge number and a Froude number, according to a millimetre sized water drop impingement. Note, that for other impingement conditions, the level of impact kinetic energy required to gain the following impingement regimes may be larger or smaller than the one described next.

The stick or float regime takes place, if the kinetic energy of the impacting droplet is very low compared to its surface energy. The drop is smoothly deposited on the free surface and may collapse into the liquid film, after some time due to gravity. The transition to the coalescence regime have experimentally been determined among others by Rodriguez and Mesler (1988), and Walzel (1980), and occurs at $We_n \cong 5$ (see Fig. 2.4b).

A moderate increase in the impacting kinetic energy leads to a different regime called the bouncing regime. Bouncing of a drop on a liquid surface occurs due to the thin air sheet that is entrapped between the drop and the liquid surfaces. The thin air sheet avoids the direct contact of both surfaces and exists during the entire bounce evolution. The abrupt deceleration, which occurs with low dissipation of energy, leads to a strong local deformation of the drop and the liquid surface. The surface tension forces, which act intense in this region, are then responsible for the drop rebound. A typical transition regime, from stick to bounce, is found for $5 < We < 10$, as reported in Rodriguez and Mesler (1988). The influence of the impact angle on the transition between the bounce and the spread have also been subject of many investigations, e.g. Schotland (1960), Zhbankova and Kolpakov (1990) or Jayaratne and Mason (1964). When the impact angle becomes flatter $\alpha < 90°$, bouncing takes place, as early as with $We < 3$. When the Weber number increases up to $We < 249$, bouncing phenomena was occasionally observed after initial coalescence and depression of the impinging surface for the impact angles of $\alpha < 14°$ by Leneweit et al. (2005). In bearing chambers, the momentum of the droplets is much higher than the one needed for the floating and bouncing droplet regime. Therefore, these two regimes play a minor role in the bearing chamber.

The spreading regime is reached for a moderate impacting kinetic energy and results in a total transfer of the drop mass and momentum to the liquid film. A cavity evolves inside the liquid film with a possible outwards moving lamella for larger incident velocities. Within the spread regime, even crown formation is possible, as long as the surface tension forces are able to avoid secondary drop generation (see Fig. 2.4c).

The most relevant drop impingement regime, which exists for large impacting kinetic energies, is the splashing of a drop. The most characteristic feature of the splashing regime is the release of secondary drops through different physical processes. Fig. 2.5 visualizes the temporal evolution of a splashing impingement onto a deep liquid pool. Immediately after the droplet touches, and before it penetrates into the liquid surface, an *ejecta* sheet shoots out from the impact region. This ejecta undergoes a characteristic instability and disintegrates in tiny drops, which have more than 10 times the impacting drop velocity. The sheet disintegration manifests with finger-like jets, protruding from the rim of the *ejecta*, or through a cylindrical liquid sheet breakup (Thoroddsen (2002) and Zhang et al. (2012)). This process is usually referred to in the literature as the prompt splashing regime.

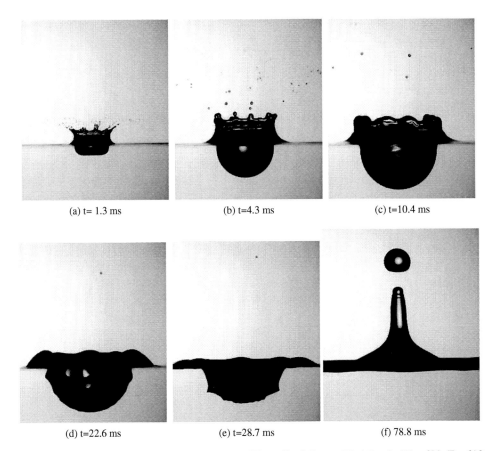

    (a) t= 1.3 ms   (b) t=4.3 ms   (c) t=10.4 ms

    (d) t=22.6 ms   (e) t=28.7 ms   (f) 78.8 ms

Fig. 2.5: Drop impact in the splashing regime: Water, D=2.8 mm, V=4.2 m/s, We=688, Fr=619, Re=14624 (images from Bisighini (2010))

At the same time, a cavity is initiated below the free surface, which has an oblate shape and a swell rises up with an abrupt corner on its side (see Fig. 2.5d). The ejecta increases the size in time and evolves to a *lamella*, which is much larger. Corrugations, which have already been existing at the rim of the ejecta now grow in magnitude, at the rim of the lamella, and build a crown like shape. Finger-like jets, which protrude then from the unstable crown during the cavity and crown expansion, are much thicker than the one in the prompt splashing regime and take usually only a fraction of the impact velocity (see Figs 2.5c and 2.5d). This sub-regime is called, on the other hand, the *corona splashing* (see Gueyffier and Zaleski (1998), Zhang et al. (2010), and Fig. 2.4d). When the corona breakup regime is finished, depending on the liquid film depth, the cavity may continue to expand until the kinetic energy of the impacting drop is converted into surface tension energy. At this point, the shape of the cavity changes from an oblate shape to a hemispherical one. While the surface tension stops the downward motion of the cavity at a certain crater depth, the sharp corner at the periphery propagates downwards and approaches the same depth as the base of the cavity. During the collapse of the cavity, the sections of the cavity periphery change their motion from outward to inward, and primarily form a polygonal shape, to a triangular shape later. The cavity afterwards retracts and reverses its direction of motion reaching the initial plane state after some oscillations.

In this subsection, a brief summary of the different impingement regimes of the drop impact dynamics have been presented. Despite the scientific importance of all the impingement regimes in the drop impact dynamics, the drop-film interaction in the bearing chambers is mostly manifested in the spreading and splashing regime. Hence, in the light of the Weber numbers, which have been identified in the bearing chambers, the stick, the bounce, and the float regime play a minor role. Additionally, from a modelling perspective, the spreading regime is easily characterized by adding simply the entire mass and momentum to the liquid wall film. However, the splashing regime requires knowledge about the secondary drop characteristics for a quantitative characterization of the mass and momentum exchange. Before the literature review dedicated to the description and modelling of the splashing products, one aspect that needs to be dealt in the drop-film interaction models is the transitional regime threshold. This threshold is usually related in terms of a combination of the impinging non-dimensional quantities. In the next subsection, the most relevant scientific publications in this topic will be discussed, by comparing the range of the impingement conditions in the bearing chambers, with the available transition thresholds in the literature. Hence, the question, whether spread or splash occurs in bearing chambers will be addressed.

2.5   Transitional Regime Thresholds between Spreading and Splashing

For the development of a drop-film interaction model, which can be applied to the impingement conditions found in the bearing chambers, it is of major importance to capture the transition between the spread and the splash of a drop impact onto a liquid surface. Since the transitional regime thresholds between the spread and the splash depend mainly on the influence of the wall, the literature review will be presented based on the influence of the wall film depth. Classifications for sub-impingement regimes depending on the liquid film depth have been suggested by Macklin and Metaxas (1976), Tropea and Marengo (1999), and Wang et al. (2002), and can principally be discriminated by: Impact on a wetted wall, ($H^* < 0.1$), a thin film, ($0.1 < H^* < 1$), a shallow/thick film, $1 < H^* < 4$, and a deep pool, $H^* > 4$. Here, the impact onto wetted walls, ($H^* < 0.1$), plays a minor role and is therefore neglected in the following review. The determination of the transitional thresholds using a combination of the dimensionless parameter have been the subject of many experimental and analytical investigations in the literature. For better comparison of the different correlations, all the thresholds are rewritten in terms of the group, $Oh^a \cdot We^b$, as proposed by Cossali et al. (1997).

Transitional Regime Threshold for the Impacts on Wetted Surfaces

For the impact onto wetted surfaces ($H^* = 0.1$), Walzel (1980) found a splashing limit of millimetre sized water/glycerin droplets onto a film made of the same liquid, with

$$K_{crit} = Oh^{-0.4}We = 2500. \qquad (2.8)$$

Yarin and Weiss (1995), on the other hand, have examined this transition regime with the impact of submillimetre, and monosized drop chains, onto wetted surfaces. Since the surface has always been wetted by the impingement of the previous drop in the chain, they defined an approximate law for the film thickness with 1/6th of the drop diameter, corresponding to the relative film height, $H^* \approx 0.1$. They proposed the splashing limit to be independent of the drop diameter and the viscosity, but to be affected by the fluid properties, the incident velocity and the impact frequency of the drop chain. The original formulation was given as

$$V_{crit} = V(\sigma/\rho)^{-0.25}/\mu_l^{0.125}f^{0.375} > 18. \qquad (2.9)$$

Samenfink (1997) has studied the transition regime for the single drop impact onto a plane wetted surface with drop sizes between $1\ mm$ and $4\ mm$. The relative film heights were defined with $H^* = 0.1$, while the surface roughness of the Perspex glass was less than $10\mu m$. Samenfink (1997) found a transition to splashing for

$$K_{crit} = Oh^{-0.4}We = 1460. \qquad (2.10)$$

By analysing in more detail the spread/crown formation and the crown formation/splash boundary, Rioboo et al. (2003) identified thresholds for

$$K_{crit,CF} = Oh^{-0.4}We = 400, \qquad (2.11)$$

and

$$K_{crit,CB} = Oh^{-0.4}We = 2100, \tag{2.12}$$

respectively. The crown breakup threshold matched the limits found by Walzel (1980), Coghe et al. (1995), and Samenfink (1997).

Vander Wal et al. (2006b) have also examined the impingement of alkane/alcohol drops on wetted surfaces. Surprisingly, the comparison to the impact onto dry surfaces showed a shift of the splashing threshold curve towards lower impact kinetic energy. Their correlation, in contrast, is depending only on the Weber number with

$$\sqrt{We} = 20, \tag{2.13}$$

and is independent of the viscosity. The shift towards lower impact kinetic energy was assessed using the kinematic discontinuity approach proposed by Yarin and Weiss (1995).

Huser (2011) studied the transition regime of monodispersed drop chains onto wetted surfaces. The impact conditions and fluid properties were selected according to the range of impingement conditions found in the bearing chambers. The splashing thresholds equalled the correlations of Coghe et al. (1995) and Rioboo et al. (2003). Fig. 2.6 additionally shows a comparison between the measurements of Huser (2011), which have been conducted in the context of this PhD thesis, and further available splashing limits for the drop impact onto wetted walls. Here, the correlation of Samenfink (1997) shows a good agreement with the splashing limit of Huser (2011). Those of Walzel (1980) and Cossali et al. (1997), over-predict the splashing initiation, which has been identified with the experimental data.

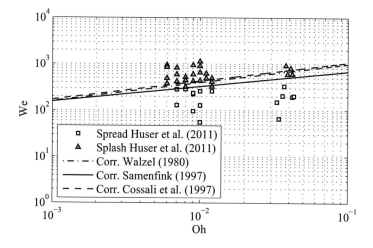

Fig. 2.6: Drop impact onto wetted walls ($H^* \approx 0.1$): Comparison of available splashing thresholds with bearing chamber impact conditions

Transitional Regime Threshold for the Impacts on Thin Liquid Films

An increase in the relative film height showed to significantly effect the border between spread and splash and particularly in the transition from a wetted wall impact, $H^* = 0.1$, to a thin film impact, $H^* < 1$. Coghe et al. (1995) studied the influence of the liquid film height on the splashing threshold using drops with sizes between $1\ mm$ and $4\ mm$, which had been generated by a syringe. With increasing film depth, a radical shift of the splashing limit towards higher non-dimensional impact energy K was found. The correlation originally contained the Weber number, the relative film height, and the Laplace number, but can be rewritten to

$$Oh^{-0.4}We = K_{crit} = 1900 + 6520 \cdot H^*. \tag{2.14}$$

Cossali et al. (1997) have conducted experiments on the single drop impacts onto thin liquid films for relative film heights, between $0.1 < H^* < 1.2$, with different water-glycerin mixtures. The non-dimensional surface roughness of the wall was given with $R_a^* = 5 \cdot 10^{-5}$. In addition to the crucial quantity, $K = We \cdot Oh^{-0.4}$, which have been identified to characterize the splashing threshold, they also proposed an equation for the splashing limit for a wide range of the Ohnesorge number Oh. With increasing relative film height and Ohnesorge number, a shift towards larger K_{crit} occurred. The correlation, which covers the splashing threshold of their experiments, is given as

$$K_{crit} = WeOh^{-0.4} = 2100 + 5880 \cdot H^*. \tag{2.15}$$

Experiments for the characterization of the splashing limit during the drop impact onto liquid surfaces have also been conducted by Kalantari (2007). He found a significant dependence on the liquid film height with a strong increase in the non-splash/splash transition towards larger critical impact energy, up to $H^* = 1$, which can be expressed as

$$K_{crit} = 1304 + 5032 \cdot H^*. \tag{2.16}$$

Coghe et al. (1995), Cossali et al. (1997) and Kalantari (2007) supported moreover the fact that as $H^* > 1$, the influence of the wall becomes less significant for single drop impact. None of them, however, included this behaviour in their proposed correlations.

Transitional Regime Threshold for Impacts on Thick Films and Deep Pools

Macklin and Metaxas (1976) has experimentally investigated the temporal crown and cavity evolution during the drop impact onto thin films, shallow and deep liquid pools, using two aqueous solutions of ethanol and glycerol. A dimensional energetic formulation was used to find the splashing threshold for different level of influences coming from the wall. Three sub-regimes using critical dimensional impact energies were found: the deep pool regime, for $H^* > 5$, the shallow liquid, for $H^* < 2$, and a transition regime in between.

Samenfink (1997) has extended his studies related to the investigation of the splashing limit by focussing more on the influence of the liquid film height for drop impingement onto thicker liquid films, $H^* > 1$. Samenfink (1997) was also able to detect the asymptotic behaviour of the splashing limit and proposed a modified correlation, which was derived from the wetted wall impacts with

$$Re = 38.1 \cdot F_{H*} \cdot (\frac{1}{Oh^2})^{0.4189}. \tag{2.17}$$

The coefficient, F_{H*}, results from a best fit to the experiments with

$$F_{H*} = 1.23 * (H^* + 0.33)^{0.19}, \tag{2.18}$$

for $0 < H^* < 3$.

Vander Wal et al. (2006a) have studied the influence of the liquid film height on the splashing threshold for the drop impact onto thick liquid films, up to deep liquid pools, $1 < H^* < 10$, using a matrix of fluids. The magnitudes of the surface tension, the viscosity and the kinematic impact conditions were systematically varied. Similarly to the previous mentioned impingement regimes, Vander Wal et al. (2006a) have observed two splashing modes, the prompt and the delayed (corona) splashing. The disappearing influence of the wall has restricted the prompt splash and inhibited the delayed splashing. Although valuable visualizations and characterization of the drop impact onto thick liquid films were studied in their investigation, no suitable correlations, which describes the observed behaviour, could be gained.

Okawa et al. (2006) have examined the normal impact of water drops onto the liquid surfaces, with drop sizes varying between $0.15\ mm$ and $4.7\ mm$, and relative film heights ranging from $0.43 < H^* < 68$. Droplet generation was achieved using a spray nozzle and a syringe. They divided the regimes into the crown formation (CF), the crown breakup (CB), the central jet formation, and the sudden ejection of secondary drops. Interestingly, they have found a splashing limit between the crown formation and the crown breakup at $K_{crit} = 2100$ over the whole range of impact conditions. Moreover, according to Rioboo et al. (2003), they also confirmed no significant influence of the liquid film depth on the threshold.

Up to this point, all the reviewed investigations have been conducted for the single drops impacting normally onto liquid surfaces. Not only in oil systems of aero-engines, but also in several other industrial applications, the oblique impact is likely to take place. Hence, its influence on the splashing threshold will be reviewed in the next paragraph.

Transitional Regime Threshold for the Oblique Impacts on Liquid Films

In order to take the influence of the impact angle into account, a different definition of the absolute impact velocity have been developed by Samenfink (1997) based on the experimental data of Schneider (1989). This modified velocity depend on the impact angle and was used for the calculation of the Weber number and the Reynolds number. The modified velocity is defined with $V_{mod} = V \cdot sin(\alpha)^{0.63}$, and is valid for the liquid film depth range of $0 < H^* < 3$. α was the angle bounded between the horizontal free surface and the absolute velocity vector of the impinging droplet.

Okawa et al. (2008) have experimentally studied the effect of the impingement angle during a water drop impact onto a plane water surface in the Weber number range, of $7 < We < 818$, at various impact angles, of $11 < \alpha < 75°$, and relative film heights, of $1.7 < H^* < 52$. In contrast to all other investigations, which had been dedicated to the influence of the impact angle on the bouncing-coalescence and the splashing threshold, they confirmed the deposition-splashing limit to be mainly dependent on the absolute impact velocity rather than on the normal component. When the normal component was used for the calculation of K, a flatter angle has resulted in a shift of the splashing limit towards lower impact energy. By replacing the normal component with the absolute velocity, a constant behaviour of $K_{crit} = 2100$, up to $\alpha = 40°$ was shown. According to their previous investigation in Okawa et al. (2006), no liquid film depth dependence on the splashing threshold was identified .

Transitional Regime Threshold for the Impacts on Flowing Liquid Wall Films

Investigations relating the influence of moving liquid films on the transition between spread and splash are very scarce in the literature. One investigation, involving drop interaction with moving films, was conducted by Samenfink (1997). The moving liquid film was generated through shear, applied by an air flow of $V = 30 \ m/s$. Chains of droplet with diameters between $100 \ \mu m$ and $200 \ \mu m$ were impacting on a wavy shear driven liquid film of a relative height of $0 < H^* < 3$. A shift of the splashing threshold curve towards lower critical impact energy, compared to steady plane liquid surfaces, were measured. The correlation was again formulated using Re and La and can be rewritten approximately with

$$Oh^{-0.4}We = K_{crit} = 696. \tag{2.19}$$

The influence of the impingement angle was well captured by using the normal component of the absolute impact velocity for the calculation of K.

A recent experimental investigation on the impingement regimes resulting from the normal droplet impact on horizontal moving films was conducted by Alghoul et al. (2011), using drop sizes between $2.47 \ mm$ and $3.86 \ mm$, and impacting on liquid films of $4.3mm < h_f < 9.4mm$. The averaged film velocity varied between $0.08 \ m/s$ and $0.2 \ m/s$. The regime boundaries have been identified using high-speed imaging from the top and side perspective. Alghoul et al. (2011) have concluded that the impact regimes were similar to those found for a static liquid film impact, whereas the transition regimes slightly differed from the one obtained by Okawa et al. (2006). Non-splash/splash boundaries were defined with $K_{crit} = 2300$ for moving liquid films.

Conclusions for the Reviewed Literature on the Transitional Regime Thresholds

Fig. 2.7 compares the different transitional regime threshold models with the detected range of impingement conditions found in the bearing chambers. It emphasizes that for the given range of

Weber numbers and Ohnesorge numbers in the bearing chambers, the spreading and splashing regime is likely to exist. Moreover, the resulting regime is decisively affected by the target surface characteristics, as indicated by the different threshold lines in Fig. 2.7.

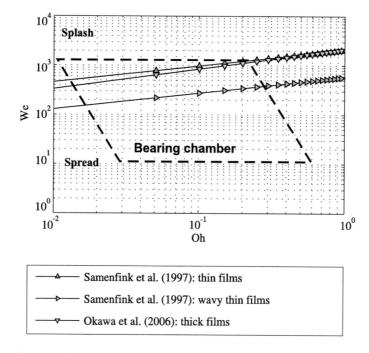

Fig. 2.7: Application of various non splash-splash correlations to the parameter range in the bearing chamber

Only a few models are covering most of the possible bearing chamber interaction conditions. An evaluation of the validity range of the available correlations in the context of the bearing chambers has concluded that the models of Cossali et al. (1997), Samenfink (1997), and Okawa et al. (2006) are applicable. Only for very high Ohnesorge numbers, $Oh > 0.141$, their applicability is challenged. The initiation of the splashing regime, for such high Ohnesorge numbers, is unlikely in the bearing chamber, since very large impact kinetic energy is required.

In conclusion and with the confidence that the splashing of droplets is a very frequent phenomenon in the bearing chambers, the next step in this chapter is the review of literature, which describes the products of splashing. A review of the literature will be presented that analyses the crown's breakup dynamics and the secondary droplet characteristics.

2.6 Secondary Droplet Generation within the Splashing Regime

The scientific literature dedicated to the description of the secondary droplet generation during drop impingement is mainly split into three parts with

- (i) theoretical studies, describing the temporal evolution of the crown and cavity, combined with the nature and growth rate of the crown instabilities, which lead to the crown breakup,

- (ii) experimental investigations, treating the disintegration process as a "black box", and deriving the correlations by evaluating the generated spray based on a variation of the impinging parameters,

- (iii) two and three dimensional numerical simulations, which directly calculate the crown and cavity evolution as well as the disintegration process.

2.6.1 Crown and Cavity Evolution in the Splashing Regime

Crown Evolution

In the very early stage of the impingement an ejecta sheet may form. Only a few authors have been able to resolve the time and length scale of this sheet expansion with the help of the experimental and numerical technique (see Weiss and Yarin (1999), Thoroddsen (2002), Zhang et al. (2012)). However, despite its scientific importance with respect to the derivation of the crown's instability, the small droplets released at this stage from the ejecta sheet are very small (1-2 μm), and carry a negligible small fraction of the total secondary droplet mass. Hence, in practical situations, this flow regime plays a minor role. The larger time and length scales, associated with the subsequent evolving *lamella*, allowed authors to investigate it much earlier and in more detail. Fig. 2.8 shows a snap shot of the lamella and crown evolving in the later stages.

Yarin and Weiss (1995) have investigated a submillimetre drop-chain impact onto a thin liquid layer and the propagation of the *lamella* and the *crown*. A square-root dependence was found for the temporal evolution of the crown diameter. The description was derived from the theory of the propagation of a kinematic discontinuity in the flow field. Yarin and Weiss (1995) proposed an analytical expression for the horizontal crown position given by

$$\frac{D_{crown}}{D} = \left(\frac{2}{3}\right)^{0.25} \frac{V^{0.5}}{D^{0.25} H^{*0.25}} (\tau - \tau_o)^{0.5}, \tag{2.20}$$

with τ expressing the non-dimensional time, τ_0 being the shifting time and V giving the impinging velocity. A shifting time is needed to adjust the starting point of the impingement.

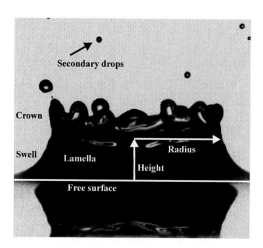

Fig. 2.8: Crown dynamics during water drop impact onto a liquid surface (Image from present PhD thesis)

Yarin and Weiss (1995) have compared their approach to the experimental data of Levin and Hobbs (1971), and found an overestimation of 10% (see, e.g. Yarin (2006)). Rieber and Frohn (1999) showed a very good agreement to the theory of Yarin and Weiss (1995) for the crown propagation by means of a remarkable three dimensional simulation using the Volume-Of-Fluid method.

Later, the theory of Yarin and Weiss (1995) was extended by Trujillo and Lee (2001) to include the viscous losses. However, they have demonstrated that the influence of the Reynolds number on the crown diameter evolution was rather small. In contrast, the crown diameter evolution, for low Reynolds numbers, was found to be considerably affected by the liquid film height.

Cossali et al. (2004) have examined the influence of the liquid film height and the incident velocity on the temporal evolution of the *lamella* and the *crown* during millimeter size water drop impact. The relative film heights ranged between $0.1 < H^* < 1.13$. Cossali et al. (2004) have stated that the crown diameter evolution was independent of the film height, but is slightly affected by the Weber number. On the other hand, the crown height, and the non-dimensional time to reach the maximum crown height, significantly depend on the Weber number. A generalized expression, for the description of the crown diameter evolution based on the one of Yarin and Weiss (1995), was proposed with

$$\frac{D_{crown}}{D} = C \left(\tau - \tau_o\right)^n .$$

(2.21)

This relation is at least valid for $\tau < 15$, with $n = 0.43$, independent of the incident velocity and the film height, and C increasing with the Weber number. The crown thickness increased with the time during the impact and was surprisingly not affected by the incident velocity and the film thickness.

The same behaviour was also observed by Fedorchenko and Wang (2004) and successfully predicted by them.

Kalantari (2007) conducted an energy and mass balance analysis in order to analytically describe the crown evolution, the crown thickness and the maximum crown height during single drop impact onto thin liquid surfaces. By taking several theoretical and empirical sub-models for the individual components within the energy balance into account, a general analytical formulation was derived

$$\sum_{n=1}^{6} A_n \cdot (H_{c_{max}}^*)^{\frac{n}{2}} = A_0. \tag{2.22}$$

Good agreement was found to the experiments of Cossali et al. (2004). The validity of the description is given for $0.25 < H^* < 1$.

Relation between the Cavity and Crown Evolution

Despite the common agreement about the theory of Yarin and Weiss (1995) describing the horizontal position of the crown, no analytical formulation exists for the vertical position of the crown. The reason for this shortcoming is due to the additional strong dependence of the crown height on different parameters, as e.g. the presence of a wall, the fluid properties, and the impinging drop parameter. However, with respect to the production of secondary drop, crown height is essential and needs to be understood.

Fig. 2.9: Visualization of the cavity for a acetic acid drop impact onto a deep liquid pool (Shadowgraph from Bisighini (2010)): We = 2177, Fr = 691, Re = 12642

Interestingly, some authors recently used the cavity evolution, during single drop impingement onto a deep liquid pool, to understand the role of the liquid film depth in thin and shallow film impacts (see Berberović et al. (2009) and Bisighini (2010)). Bisighini (2010)) has shown that a drop impingement onto a thin liquid layer can be described by using an equivalent drop impact onto a deep liquid pool. Here, if the maximum cavity depth, $\Delta_{max,deep}$, for the equivalent deep pool impact, is larger than the relative film height for the thin film impingement, larger crown heights were visible with an enhanced secondary droplet production. This effect becomes weak when the maximum cavity depth, for an equivalent deep pool impingement, is smaller than the relative film height. However, this relation was given only qualitatively with the help of shadowgraphs. They have additionally proposed a new classification for the role of the liquid film depth in single drop impingement on liquid surfaces using the maximum cavity depth for deep pool impacts with (i) thin Films, $H^* <= \frac{1}{10}\Delta_{max,deep}$, (ii) thick films, $\frac{1}{10}\Delta_{max,deep} < H^* < \frac{3}{2}\Delta_{max,deep}$, and (iii) deep pools, $H^* > \frac{3}{2}\Delta_{max,deep}$. This classification, together with the maximum cavity depth for an equivalent deep pool impact, will also be used in the present thesis to quantify the influence of the liquid film depth on the secondary drops generation. The maximum cavity depth, for an equivalent deep pool impact, is therefore chosen, since it delivers a good measure for the fraction of kinetic energy that is transferred to the liquid wall film during a single drop impingement. Hence, in order to classify the impingements with respect to the initial film height, the maximum cavity depth for deep pool impingement needs to be determined. Bisighini et al. (2010) has found an analytical expression for the temporal evolution of the cavity depth by dividing the impact evolution into two stages. The initial stage takes place for $\tau < 2D/V$ and is characterized by the initial drop deformation and the crater formation. In the later stages, for $\tau > D/V$, the inertial flow in the liquid pool is balanced by the surface tension and the gravity. Similar to the results obtained by Fedorchenko and Wang (2004) and Berberović et al. (2009), the initial stage is only dependent on the non-dimensional time but with a slightly different calculated coefficient of $\Delta \approx 0.44\tau$. For the later stages, the flow around the crater is approximated by potential flow field, past a moving and expanding sphere, and is obtained through a balance of stresses at the crater interface. The cavity depth is defined in their model as a combination of the axial coordinate of the centre of the expanding sphere and the crater radius. The numerical solution of the following second order ordinary differential equation, including the viscous, the surface tension and the gravitational effects, describes the temporal evolution of the crater depth with

$$\ddot{\alpha} = -\frac{3}{2}\frac{\dot{\alpha}^2}{\alpha} - \frac{2}{We\,\alpha^2} - \frac{1}{Fr}\frac{\zeta}{\alpha} + \frac{7}{4}\frac{\dot{\zeta}^2}{\alpha} - \frac{4}{Re}\frac{\dot{\alpha}}{\alpha^2}, \tag{2.23}$$

and

$$\ddot{\zeta} = -3\frac{\dot{\alpha}\dot{\zeta}}{\alpha} - \frac{9}{2}\frac{\dot{\zeta}}{\alpha} - \frac{2}{Fr} - \frac{12}{Re}\frac{\dot{\zeta}}{\alpha^2}. \tag{2.24}$$

Here, α represents the crater radius, and ζ the axial coordinate of the centre of the sphere. Then, the cavity depth results by the sum of α and ζ with $\Delta = \alpha + \zeta$.

The initial conditions were extracted from their experimental data at $\tau^* = 0$, and $\tau^* = 2.3$, leading to the following numerical values $\alpha = 1.12$, $\dot{\alpha} = 0.17$, $\zeta = 0$, and $\dot{\zeta} = 0.27$. The solution of these ODEs also allowed the prediction of the maximum cavity depth after drop impact onto a deep liquid pool.

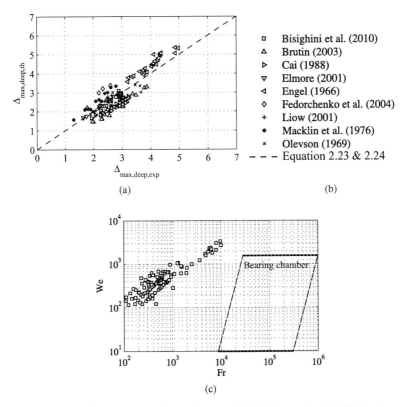

Fig. 2.10: Validity of the theoretical formulation of Bisighini et al. (2010) for the maximum cavity depth through comparison with experimental results in the literature and the parameter range in the bearing chambers

The author has stated that an extrapolation towards more engine relevant impact conditions with large Weber numbers, Froude numbers, and low Capillary numbers, would be allowed with this theory and proposed a $We - Fr$ map, showing the maximum cavity depth as iso-lines. The validity of the analytical model of Bisighini et al. (2010) is displayed in Fig. 2.10a. Here, it is clearly shown that a reasonable agreement between the theoretical model and the experimental data of Olevson et al. (1969), Macklin and Metaxas (1976), Liow (2001), Fedorchenko and Wang (2004), Engel (1966), Cai et al. (1988), Brutin (2003), and Bisighini (2010) exists. Fig. 2.10c shows, moreover, a comparison between the experimental impingement conditions, derived from the previous scientific publications and the range of impingement conditions, found in bearing chambers, in a $We - Fr$ map. Here, a clear lack of the measurement data for the large Froude numbers is identified. This gap arises from the fact that most of the investigations reported here have used millimetre sized droplets with large gravitational forces acting during the impingement in order to describe the cavity dynamics. For the droplet impingement in the submillimeter range, the gravitational forces acting during the impingement are negligible compared to the inertial forces of the droplet, which is equivalent to very large Froude numbers. The experimental

description of the cavity dynamics for these small time and length scales is still very difficult
with available high-speed imaging techniques.

2.6.2   Theory of Instability Driving the Break up of the Crown's Rim

Some authors used the analytical description of the crown propagation, discussed in the previous
section, and adopted an instability theory in order to predict the number and the size of the
secondary droplets released in the splashing regime. For completeness, a selection of the most
relevant studies is given in this section.

The extension of the *lamella* is followed by the formation of a rim at the edge that is subject to
surface tension and fed with liquid by the *lamella* sheet, while the rim is further extending.

Fig. 2.11: Crown disintegration during the impact of a silicone oil droplet (Re=966, We=874,
$H^* = 0.2$). The image is taken from the publication of Zhang et al. (2010).

A symmetry-breaking corrugation leads later, in the non-linear phase of the instability, to cusps
where finger-like jets protrude which break up into secondary drops. In Fig. 2.11, the shapes
of the splashing crown, the finger-like jets and the secondary drops for a silicone oil droplet
impact onto a liquid surface are depicted. There are various theories cited in scientific literature
describing the nature of the instability driving the sheet breakup at the rim of the crown. Before
reviewing the scientific literature, a brief description of the most relevant and agreed instability
mechanisms driving the breakup of the crown's rim in the drop impingement, the so-called
Rayleigh-Plateau and the Rayleigh-Taylor instability, will be given.

Rayleigh-Plateau Instability

Plateau (1873) observed first that if a cylindrical jet length becomes larger than its circumference, a characteristic instability evolving on the sheet was always visible. The mathematical description of this instability was introduced later by Rayleigh (1878) by means of a linear instability analysis, applied to viscous and inviscid water jets. Here, a capillary driven breakup of a cylindrical fluid sheet subject to surface tension was described (Rayleigh (1892)). The result of the linear instability analysis delivered a dispersion relation where instability of the liquid sheet occurs only if

$$kR_{cyl} < 1, \tag{2.25}$$

where $k = \frac{2\pi}{\lambda}$ is the wave number and R_{cyl} is the initial radius of the cylinder. The component of the disturbance that grows the fastest is given with $kR_{cyl} \approx 0.697$.

Hence, the resulting peak wavelength λ_{max} of the perturbation that has the highest growth rate, corresponds to 4.5 times the radius of the cylindrical sheet R_{cyl} with

$$\lambda_{max} = 4.5 \cdot R_{cyl}. \tag{2.26}$$

Accordingly, the secondary drop diameter results then to 3.78 times the radius of the cylinder,

$$D_{sec} = 3.78 \cdot R_{cyl}. \tag{2.27}$$

Figure 2.12 shows two images recorded by Zhang et al. (2010) depicting the crown's rim produced by a silicone drop impingement onto a liquid surface of $H^* = 0.2$ at two different time instances, at $t = 1.85\ ms$ and $3.15\ ms$, respectively. The images express clearly that there is a clear relation between the number of the secondary drops released at $t = 3.15\ ms$ and the peak wave length of the corrugated rim at $t = 1.85\ ms$. Moreover, the size of the secondary drops is clearly a multiple

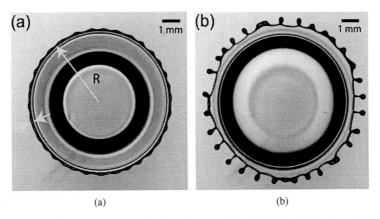

(a)   (b)

Fig. 2.12: Crown splash during a single drop impact (Re=894, We=722, $H^* = 0.2$). Relation between the Rayleigh-Plateau instability wavelength and the number of secondary droplets. Images are taken from the publication of Zhang et al. (2010) at $t = 1.85\ ms$ and $t = 3.15\ ms$ after impact.

of the crown's rim diameter. They theoretically proved that for the underlying impingement conditions, specifically at $Re = 894$ and $We = 722$, the Rayleigh-Plateau instability was driving the breakup of the crown's rim.

Rayleigh-Taylor Instability

A further mechanism describing the growth of perturbations at the free surface between two or more phases is given by the Rayleigh-Taylor instability. The Rayleigh-Taylor instability sees beside the pure capillary amplification of the disturbances acting at the free surface an additional amplification arising from the deceleration of a fluid surface relative to a less dense external medium. The deceleration of the fluid surface may last until it reaches the Taylor-Culick velocity (see Taylor (1959)), when viscous effects are small and surface tension forces dominant; the rim deformation is then initiated and driven by the Rayleigh-Taylor instability. The Rayleigh-Taylor instability is hence a time-dependent instability. Assuming two incompressible fluids with the liquid density much larger than the gas density, and surface tension forces acting at the free surface, a characteristic length scale driving this instability results with

$$l_a = \sqrt{\frac{\sigma}{\rho a}}. \qquad (2.28)$$

Here, a corresponds to the deceleration of the free surface (see Zhang et al. (2010)). According to the results of the linear stability analysis in the previous section, a dispersion relation can also be mathematically developed for the present instability. Here, disturbances acting at the rim and subject to surface tension grow if

$$k R_{cyl} < \frac{1}{l_a}. \qquad (2.29)$$

The wave number (maximum instability wave length) of the fastest growth rate is slightly smaller than $1/l_a$ but larger than the one evaluated for the Rayleigh-Plateau instability (see Zhang et al. (2010)).

Literature Review of the Instability Driving the Crown's Breakup

Almost all available observations of this process agree that the generation of secondary drops evolve from the disintegration of finger-like jets protruding from cusps formed at the rim of the *lamella* (see, e.g. Levin and Hobbs (1971), Stow and Hadfield (1981), Macklin and Metaxas (1976), Yarin and Weiss (1995), Cossali et al. (1997), Vander Wal et al. (2006a), and Zhang et al. (2010)). Despite this agreement, theories describing the formation of the cusps, the fingers and the secondary drops are numerous and still not completely satisfying. The small time and length scales occurring during the splash, particularly in the earliest stages of the *ejecta* expansion, make a quantitative description of the origin of the extremely small perturbation at the rim really challenging.

Worthington (1879) was one of the first authors to suspect that the main instability, which would cause secondary drops from the crown rim, was due to the Rayleigh-Plateau instability, which causes cylindrical, toroidal jets to break up as described by Rayleigh (1878). Unfortunately, the lack of measurement techniques for detecting the number and size of secondary droplets made a direct comparison impossible at that time.

Rieber and Frohn (1999) have numerically investigated the rim instability occurring at the rim of the splashing drop impacts onto thin liquid layers. They solved the Navier-Stokes equations for an incompressible fluid in three dimensions and used the Volume-Of-Fluid method. Only by introducing strong initial perturbations, specifically a random disturbance based on a Gaussian distribution, to the initial velocities of the film and the droplets, did they manage to obtain a splashing mechanism that was close to experimental observations. Hence, they concluded that the Rayleigh instability was driving only the formation of cusps in the later stages of the splashing process but not at the early stages. Bremond and Villermaux (2006) also supported this mechanism for the breakup of the *lamella* produced after an oblique jet impact.

Yarin and Weiss (1995) have compared Rayleigh's theory to theirs and to the experimental data of Levin and Hobbs (1971), and found a discrepancy in the amount of secondary drops breaking up simultaneously from the crown top.

With this, Yarin and Weiss (1995) have proposed a non-linear amplification mechanism of small perturbations at the rim due to roughness, or rippling, which propagates in the very early phase of the impact.

Gueyffier and Zaleski (1998), and Fullana and Zaleski (1999), proposed another mechanism for the formation of jets, in the earliest stage of the impact evolution, driven by the sudden acceleration of the interface (Richtmyer (1960)). They were able to produce the fingering phenomena by imposing a small harmonic perturbation on the drop shape and a uniform velocity profile.

Roisman et al. (2006) investigated the transverse stability of a free rim bounding a free liquid sheet which is generated during the impact of a single droplet onto a liquid film. They proved that this type of instability is responsible for the cusp formation and the protrusion of finger-like jets from the crown's rim. With the help of a viscous linear analysis of the transverse instability, Roisman et al. (2006) showed that the flow feeding the rim from the lamella sheet stabilizes the rim. In contrast, the rim deceleration acts more as an amplifier of the rim perturbations. However, the wave length of the fastest growing mode was in good comparison with the capillary driven instability of Rayleigh-Plateau and only slightly affected by the rim acceleration. The breakup of the crown's rim could hence be described by a combination of the Rayleigh-Taylor instability in the early stages of the crown's splash whereas by the Rayleigh-Plateau instability when rim deceleration was very small. The crown's splash mechanism found by Roisman et al. (2006) was also confirmed later by Agbaglah et al. (2013) and Agbaglah and Deegan (2014).

A remarkable investigation on the water or ethanol drop fragmentation mechanism during impact onto solid surfaces was conducted by Villermaux and Bossa (2011). The dynamics governing the radius of the *lamella*, its stability and the resulting fragment drop size distribution was experimentally documented. A break up after the Rayleigh-Taylor mechanism was identified for the disintegration of the rim into secondary drops. They identified specifically a body force pushing the fluid radially outwards from the rim that was responsible for the induction of a breakup after the Rayleigh-Taylor mechanism.

In conclusion, the variety of the investigations and theories, available for the derivation and nature of the crown instabilities confirm that the physics are still not fully understood. In the present investigation, these theories are used to derive the nature of instability driving the crown break up in the direct simulation of the splashing crown. The following research activities chose a more empirical approach by going the way around the physics of the crown's breakup and studying the effect of several influence parameters on the resulting post-impingement secondary drop characteristics. This will be shown in the next section.

2.6.3   Secondary Drop Characteristics during the Drop Impact

The missing link between the origin of perturbation and the production of secondary drops lead to the fact that various researchers attempted to correlate the post-impingement products, e.g. the size, velocity, and ejected mass fraction of the secondary drops with the impingement conditions.

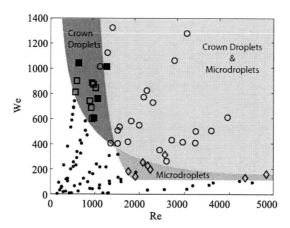

Fig. 2.13: Secondary droplet produced from the prompt and the corona splashing regime during single drop impingement onto a liquid layer of $H^* = 0.2$. Diagram from Zhang et al. (2010). No splash is represented by the black small circles, crown droplets with and without microdroplets by the open circles, and microdroplets without crown droplets by the diamonds. The filled squares indicate the parameter set for all experiments at $H^* = 0.2$ conducted in Zhang et al. (2010).

Little attention has been paid to the hydrodynamics of the drop deformation occurring in-between. Fig. 2.13 depicts a map, developed by Zhang et al. (2010), showing the splash morphology for the various impacting Weber and Reynolds numbers. Here, three different secondary drop patterns produced during a single drop impingement were defined. For $Re < 1500$ and $We > 300$, only the secondary droplets produced from the corona breakup were identified. On the other hand, for $We < 300$ and $Re > 1500$, only secondary droplets in the prompt splashing regime were found. Outside of these boundaries the authors found both secondary drop patterns to occur. Levin and Hobbs (1971) have studied the products of splashing during the single drop impacts onto solid and wetted surfaces. They measured, as one of the first contributions, the size distribution of the secondary drops produced by a $3\ mm$ water drop at a velocity of $4.2\ m/s$ on a copper hemisphere of the diameter $2.5\ cm$. A log-normal size distribution represented the measured sizes well with diameter peaks of the probability density functions between $100\mu m$ and $150\mu m$.

Stow and Stainer (1977) have conducted extensive experiments of the products of splashing during millimetre sized drop impacts onto dry and wetted walls. The number and sizes of the secondary drops were analysed with respect to several parameters of influence, e.g. the impact velocity, the drop diameter, the surface tension, the target surface curvature and roughness, and the depth of the liquid film. The measurement technique adopted was based on a photographic emulsion process on a prepared surface positioned near the impact region. The surface was treated by an emulsion in order to detect the sizes and number of secondary drops. Here, the sizes were proportional to the ionized darker regions of the prepared surface. Secondary drops below $70\ \mu m$ could not be detected by this measurement technique. Stow and Stainer (1977) showed the number of secondary drops to increase with the surface roughness, the impact velocity, and the drop size, but decrease with an increase in liquid film depth. A surprising result was the reduction of the number of the secondary drops with a decrease in the surface tension. However, the use of detergents to decrease the surface tension of a fluid is not always suitable, when the distribution of the detergent near the free surface changes with the impacting forces in particular. This measurement uncertainty could explain the unexpected results obtained by Stow and Stainer (1977). Their experimentally determined secondary drop size distribution could also be well fitted by log-normal density distribution functions as in Levin and Hobbs (1971). Moreover, the mean secondary drop sizes increased with the impacting drop size, the surface roughness, and the depth of the liquid film. In contrast, a decrease of the mean secondary drop sizes occurred only with increasing impact velocity and a reduction of the surface tension.

From the experimental investigation of Mundo et al. (1995), a first quantitative description of the secondary drop characteristics following from drop chain impacts onto a rotating dry, cold aluminium disk, was proposed. The disk rotation was needed to avoid interference between the subsequent impacts and vary the impingement angles. Although a dry surface impact was stated, the authors confirmed later that a thin liquid film may be present due to the very large roughness of the disk surface (see Mundo et al. (1997)). A large parameter variation was investigated with the change in impacting drop sizes, velocities, the fluid properties and the impingement angles. Influence of these parameters on the secondary drop sizes, velocities and ejected mass was studied using a phase Doppler anemometer. They concluded that the drop size distribution became narrower with smaller mean sizes for an increasing K- number. The velocity distribution of the secondary drops resulted in larger values with an increase in impact velocity and steeper ejection for flatter impingement angles. They proposed empirical correlations describing the secondary drop characteristics and ejected mass fraction using solely the K number.

Yarin and Weiss (1995) looked also at the secondary drop characteristics and ejected mass fraction during the impact of a submillimeter drop chain onto solid surfaces (covered by a thin liquid layer). The number and sizes of the secondary drops were measured using an image processing technique of the CCD camera recordings. The secondary drop sizes were found to range from a few microns up to one quarter of the impacting drop sizes which ranged between $70\mu m$ and $340\mu m$. The peak of the size distribution occurred at $D_{sec}/D = 0.06$ for all the impact velocities investigated. The total ejected mass fraction was obtained by calculating the measured volumetric mean secondary drop size and number of the secondary drops. The ejected mass fraction increased with impact velocity, having a maximum at $V \approx 22m/s$, and decreased again for larger values of the incident velocity.

The impact of water drop chains onto shear-driven thin liquid water films was extensively investigated by Samenfink (1997) and Samenfink et al. (1999). In their experiments, the shear acting on a thin liquid film, which was induced from an airflow of $30\ m/s$, provided a wavy film. Here, the drops, dispersed from a chain due to turbulence fluctuations, were impacting. The sizes of the impacting droplets ranged between $100\mu m$ and $210\mu m$ with velocities between $9.1\ m/s$ and $23\ m/s$. The averaged relative film height and the impingement angles were also varied over a wide range, $0 < H^* < 3$, and $25° < \alpha < 90°$. The secondary drop characteristics were measured using a phase Doppler anemometer, while the deposited mass fraction and the composition ratio were extracted by a new integral method, based on the conductivity and volume flux measurements. Samenfink (1997) stated that the waviness of the thin liquid film appeared not to affect the secondary drop sizes, velocity, ejection angle and the deposited mass fraction. As expected, the ejected mass fraction increased with increasing impact kinetic energy, but decreased with a reduction of the averaged liquid film height. The changes in the value of the impact angle showed a maximum of the ejected mass fraction at $\alpha = 60°$. From the normal impact up to this angle the ejected mass increased while it did decrease beyond $\alpha = 60°$. The secondary drop sizes did again well fit to a log-normal distribution function and seemed to be significantly affected by the impact energy and the impingement angle, whereas only little affected by the relative film height. Similar to Mundo et al. (1995), an empirical model was proposed by Samenfink et al. (1999) with correlations for the secondary drop characteristics, the ejected mass fraction and the composition ratio with the help of the Reynolds and Laplace number.

Okawa et al. (2006), and Okawa et al. (2008), have studied the ejected mass fraction by measuring the number and sizes of the secondary drops during normal and oblique water drop impact onto a liquid surface with relative film heights ranging from $H^* = 0.0275$ up to $H^* = 68$. The drop sizes varied over a wide range, from $120\mu m$ to $4.7\ mm$, by using a syringe and a spray nozzle. Impingement angles were also changed between $15° < \alpha < 80°$. By increasing the non-dimensional impact momentum, the number of secondary drops increased exponentially, while slightly decreasing with increasing liquid film height. The size distribution in contrast appeared to be independent on the non-dimensional momentum K, and the relative film height, H^*, and constant around $D_{sec}/D = 0.1$. Accordingly, the ejected mass fraction was only affected by the number of the secondary drops released from the crown. A variation of the impingement angle, between $\alpha = 40 - 80°$, indicated a significant increase in the secondary drop size, while the number of ejected drops appeared to be independent on α. As a consequence, the ejected mass fraction increased according to the increase in drop sizes. For flatter impact angles, $\alpha = 20 - 40°$, the secondary drop size appeared to be reduced slightly, whereas the number of secondary drops dramatically decreased resulting in a decrease of the ejected mass fraction. The latter behaviour seemed to suite well with the predictions of Samenfink (1997) in the last section.

Experimental Limitations of the Reviewed Investigations

The review of literature dedicated to the characterization of the secondary droplet generation proposed a brief overview of the available experimental single drop impingement data, where most of the correlations and drop-film interaction models in the literature have been derived. For this reason, it is essential to evaluate not only the validity range of the experimental investigations in the light of impingement conditions close to the bearing chambers, but also the adopted measurement techniques.

The Validity Range

A very scarce amount of secondary drop data particularly for single drop impingement onto thick liquid film and deep liquid pools, as they occur in the bearing chambers, have been identified.

The adopted test fluids were either not varied at all by using only water or only changed over a narrow range, leading to a lack of generality for the established secondary drop characteristics. The effect of the fluid properties, for the conditions identified in the bearing chambers in particular, has therefore not been sufficiently understood.

Another aspect that is not taken into account in the available literature, is the lack of suitable single droplet generation methods, ensuring a controlled production of the drop sizes in the submillimetre range with large velocities. This constraint caused many researchers to use either simple syringe producing large millimetre sized drops or drop-chain generators with high-frequency drop impingement rates. With millimetre size drops, it is possible to reach large impacting kinetic energies with relatively low incident velocities. Hence, the splashing regime is entered with relatively low effort. This makes this droplet generation source still very attractive in research. However, the impinging Froude numbers with such generic drop generators significantly vary from the ones in practical spray impingements as occurring in the bearing chambers. The question whether the secondary sprays produced from large millimetre sized drop impacts can also describe submillimeter drop impingement is yet not addressed in the literature, but will be answered with the results of the present thesis. On the other hand, drop-chain generators are known to produce droplet impingements with 20000 events per second. This implies that each drop impingement event cannot be evaluated in isolation from the other events since pre- and post-impingement events are interacting with each other.

The Measurement Technique

With respect to the measurement techniques, either laser optical methods or imaging techniques have been applied. One of the major advantages of the imaging techniques, combined with particle sizing, is the correct determination of (non spherical) droplets that are known to be decisive for the ejected mass calculation. Limitations in power of stroboscopes and continuous light sources in conjunction with restricted pixel count, exposure times and frame rates of the high-speed recording device make the determination of the very small drop sizes with high velocities very challenging. The accuracy of the drop size measurements even for large drop sizes is strongly depending on the position of the droplets in the depth-of-field and the diffraction sensitivity of the recording device. Large uncertainties may hence occur for the mass flux measurement of the secondary sprays if the evaluated secondary droplet is outside the depth-of-field. None of the reviewed literature using imaging technique referred to the uncertainty of drop size measurement arising from a drop outside the depth-of-field.

The application of laser optical methods in the drop impact dynamics eliminates most of the disadvantages arising from imaging techniques. Several authors, as e.g. Mundo et al. (1995), Samenfink et al. (1999), Roisman and Tropea (2001), and Kalantari (2007), successfully applied laser optical techniques for the characterization of the secondary sprays. The main issue that comes with the application of phase Doppler anemometer is the quantification of the ejected mass fraction carried by the secondary drops during the drop impact. Experimental investigations relying on the mass flux measurement from a phase Doppler anemometer are known to deliver inaccurate results, due to the fact that measurement of non-spherical drops contains high uncertainties (see Dullenkopf et al. (1998)). The non-spherical droplets are mostly represented by the large sizes in the secondary drop size distribution, where the gravitational forces cause long-lasting drop oscillations. Hence, mass flux measurements using PDA techniques remain questionable without an accurate detection of these drop sizes. Roisman and Tropea (2001) and Muehlbauer et al. (2010) identified furthermore the shortcoming of conservative PDA techniques related to a correct quantification of the measurement volume and count errors due to multiple particles in the measurement volume. They proposed for both shortcomings corrections that showed indeed improvement of the mass flux measurement in comparison with the patternator results for different spray types.

All of the difficulties and shortcomings in experimentally accessing the products of splashing, together with increasing computational resources, and improving multiphase flow modelling techniques, motivated different authors to attempt the direct simulation of a splashing impact by means of a detailed numerical method. An advantage of the numerical simulation compared to the experimental investigation is the definition of well-defined boundary conditions, as e.g. the sphericity of the impacting drops or the isolated variation of the impacting drop parameters. On the other hand, all the quantities of the flow field can be accessed and analysed without major complications. The next section analyses the most relevant numerical methods applied for the direct simulation of a drop impingement.

2.7   Numerical Simulation of the Single Drop Impact onto Liquid Films

In this section, an overview of the available numerical investigations related to the drop impact dynamics, using the various numerical methods, is presented. Due to the large amount of the available literature on the different aspects of the drop impact dynamics, only the relevant publications for this investigation are to be discussed. For more details on the numerical simulation of the drop impact onto dry and heated surfaces, further publications as, e.g. of Pasandideh-Fard et al. (1996), Bussmann et al. (1999), Bussmann et al. (2000), Geldorp et al. (2000), Eggers et al. (2010), Mandre and Brenner (2012), and Yokoi (2013), can be consulted.

Gueyffier and Zaleski (1998) have numerically examined the impact of a droplet on a liquid surface, solving the Navier-Stokes equations coupled with a Volume-Of-Fluid method on a staggered MAC-type grid. Axisymmetric solutions of the numerical method showed a good agreement with the theoretical description of Yarin and Weiss (1995) for the crown diameter evolution. The two-dimensional simulations demonstrated large pressure maxima at the first contact between the drop and the target surface. By applying mirror boundary conditions, one quarter of the complete flow field was simulated with the three dimensional method. The fingering phenomenon was captured by the numerical method, only when small harmonic perturbations were imposed on the impacting drop and occasionally even without being affected by surface tension and viscosity. A possible instability mechanism was found similar to the one proposed by Richtmyer (1960).

Rieber and Frohn (1999) have presented three dimensional numerical simulations of a single drop impingement onto thin liquid layers for different impacting Weber numbers. The numerical method was also based on the solution of the Navier-Stokes equations, coupled with the Volume-Of-Fluid method on a MAC-grid. Similar to Gueyffier and Zaleski (1998), they have simulated one quarter of the full problem, assuming symmetry and periodic boundary conditions, with a Cartesian grid, which contained approximately 32 million computational cells. The disintegration process of the crown rim appeared to be very similar to the experimental observations. This was also confirmed by a good agreement of the crown diameter evolution with the theory of Yarin and Weiss (1995). The fingering process of the crown's rim was, however, obtained only by imposing perturbations on the drop and the target surface using Gaussian distributions.

Josserand and Zaleski (2003) have conducted a two-dimensional numerical and theoretical investigation of the single drop impact onto a thin liquid layer using VOF and the incompressible

Navier-Stokes equations. Josserand and Zaleski (2003) concentrated particularly on the neck region after the contact between the drop and the target fluid and proposed a theory predicting the transition between splash and spread. This criterion was formulated with the help of the Weber and Reynolds numbers and is proportional to a critical value depending on the flow-field inside the jet. Their splashing threshold appeared to be in good agreement with their numerical and experimental results.

Nikolopoulos et al. (2007) was one of the first authors to develop an enhanced numerical methodology based on the solution of the Navier-Stokes equations and the VOF method, allowing the three-dimensional simulation of a droplet impinging onto a wall film. The computational effort was reduced by introducing a dynamic Adaptive Mesh Refinement (AMR) technique, which refined the mesh at locations where it was needed while leaving the other regions coarse. The enhanced methodology allowed to calculate the complete three-dimensional flow field with lower computational effort and an increased level of accuracy at the same time. Three test cases were selected with different Weber numbers from which two were identical to those chosen by Rieber and Frohn (1999). The model appeared to predict the disintegration process for the higher Weber numbers with acceptable accuracy. However, for lower Weber numbers, a disintegration mechanism took place with a detachment of the *lamella* rim from the liquid sheet and a breakup driven by the capillary forces. Since such a crown breakup procedure could not experimentally be confirmed in the existing literature, it is questionable, whether this process is new or due to a numerical inconsistency. Interestingly, in contrast to the numerical method of Rieber and Frohn (1999), Gueyffier and Zaleski (1998) and according to Peduto et al. (2011b), no artificial perturbations have been imposed to obtain the correct shapes of a splashing impingement.

A coupled Level Set and Volume-Of-Fluid method (CLSVOF) was developed by Yokoi (2008) and applied to drop impact on a thin liquid layer. The numerical method was first verified with several fundamental test cases, as e.g. the Rayleigh-Taylor instability. Then, the three-dimensional numerical method was adopted for the simulation of two single drop impact scenarios with various Weber numbers in the splashing regime. The numerical results appeared very promising, although not resolving the expected fingering disintegration process, but similar to Nikolopoulos et al. (2007) with a detaching cylindrical sheet breakup later.

Gomaa et al. (2009) have studied the physical process of a drop impact on a thin liquid film using a VOF based code (FS3D) for kinetic impact energies close to the splashing limit. Their main objective was to examine the influence of the grid resolution on the process of splashing. Four different three-dimensional simulations with a grid refinement of two in each dimension were examined. A quarter of the complete flow field was calculated for each dimension with periodic boundary conditions at their periphery. The maximum grid size comprised about 135 million computational cells. The breakup process of the crown's rim was initiated only by introducing a random undirected noise field characterized by the standard deviation of the initial impact velocity.

By comparing the interface topology for the various simulations at similar time-steps, they denoted a significant grid influence that diminished when adapting the time-steps to the time-scales of the physical process of splashing.

An advanced free surface axisymmetric capturing model based on a two-fluid formulation of the classical volume-of-fluid (VOF) approach was applied by Berberović et al. (2009) for the cavity expansion during drop impact onto liquid layer of finite thickness. The advanced model obtained a sharper interface by introducing an additional convective term into the transport equation. The two-dimensional and axisymmetric simulations were validated with the experimental data of van Hinsberg (2010) and showed very good agreement. Better insights into the flow field of the cavity were gained that explained for instance the formation and propagation of the capillary waves along the cavity surface. In particular, the results describing the flow field near the bottom wall, helped to formulate scaling relations for the residual film thickness.

Based on the numerical methodology developed in Gomaa et al. (2009), Gomaa et al. (2011) have conducted a parameter study on Diesel or oil drop impact onto wetted surfaces for Diesel engine relevant conditions. The grid resolution was kept in a way to allow an extensive parameter variation with respect to the computational time, specifically around two million cells. The numerical simulations were first validated for millimetre size water drop impact onto wetted surfaces of the same fluid. A qualitative comparison proved that the splashing topology was not resolved adequately due to the inappropriate grid resolution. However, the secondary drop number and sizes gave good comparison thereby leading the author to the conclusion that the numerical method was applicable for a parameter study at Diesel engine relevant conditions. Around fifty simulations were carried out with variation of the impact velocity, the impact diameter, the fluid properties and liquid film height. Based on the developed data base a model has been proposed containing the splashing limit, the kinematic properties and the ejected mass carried by the secondary drops.

Fest-Santini et al. (2011) and Fest-Santini et al. (2012) applied also 2D axisymmetric and 3D simulations with a similar numerical method as Berberović et al. (2009) to drop impacts onto deep liquid pools. The 2D method was applied to impact conditions were the wall played a minor role with a good comparison to high-speed imaging results and temporal cavity evolution. Surprisingly, they denoted better prediction of the 2D method than with the 3D method and justified it with an inappropriate 3D mesh resolution. Fest-Santini et al. (2012) stated furthermore that the fluid trampoline regime could not be resolved with both approaches.

Beside the mesh based methods, there are moreover various promising mesh-less approaches available in the literature that need to be emphasized, e.g. the Smoothed Particle Hydrodynamics (SPH) (see Hoefler et al. (2013)). Originally this method was applied to astrophysical processes, as e.g. galaxy generation and star collisions by Gingold and Monaghan (1977). The main advantage of this method particularly for the application to multi-phase flows is the direct transport of the free surface based on the Lagrangian formulation of the transport equations despite the extensive modelling required in mesh-based approaches. Two of the decisive challenges of this method are the implementation of appropriate boundary conditions and modelling of phase interactions, as e.g. the liquid-gas and the liquid-wall interaction (see Wieth et al. (2014) and Hoefler et al. (2012)). It is obvious that in astrophysics the boundaries are often far away from the interesting local phenomena thereby not affecting its evolution. The fluid mechanics in practical situations, however, are mainly controlled by the boundary conditions: they need therefore to be reflected by the method. For this purpose, several research activities are dealing with the implementation of appropriate boundary conditions. Zhang et al. (2008) applied smoothed particle hydrodynamics to the spreading, splashing and solidification process during drop impact onto an inclined dry surface. The effect of impingement angle and temperature on the splashing limit was successfully simulated and compared to literature values. The mesh-less numerical method proved to be promising for typical drop impact dynamics scenarios but need still additional work to reach the level of accuracy of mesh-based approaches. Further very promising investigations, using SPH, were developed and applied by Hoefler et al. (2013) to primary atomization for gas turbine fuel preparation and to typical bearing chamber free surface flows by Kruisbrink et al. (2011), respectively.

Limitations of the Reviewed Numerical Investigations

The reviewed literature of numerical simulations has clarified that various methods may have the ability to directly simulate the drop impact dynamics. However, some of them, as the coupled CLS-VOF method or the SPH are still in their development stage and not fully established for an accurate prediction of the disintegration process during drop impact. Research activities using the Coupled Level-Set and Volume-Of-Fluid method (CLSVOF), for instance, are still dealing with issues arising from mass imbalances near the free surface. In SPH methods, on the other hand, modelling surface tension appeared not to be trivial and needs still some further provisions and time for full establishment. The Volume-Of-Fluid method, adopting a homogeneous eulerian approach has been proven to have a very good potential and to be more suited for the direct simulation of the drop impact dynamics. However, a full direct simulation of spray impacts remains still an insurmountable target with the given computational resources. The application of the Volume-Of-Fluid method for the direct simulation of the single drop impacts is proven very accurate for the cavity dynamics using a two-dimensional, axisymmetric method, and delivered good agreement with the experimental results. However, even the two-dimensional simulations contained large number of computational cells limiting extensive parameter studies which are needed to get a detailed understanding of the physics. Three dimensional simulations were also conducted in the literature to directly simulate the crown expansion, the crown's disintegration and the secondary drop propagation during single drop impact. Since the number of elements in these investigations increased up to several hundred millions, computational time and effort was reduced in the majority of the investigations by assuming symmetry boundary conditions and calculating only a quarter spatial fraction of the entire drop impingement process. Such periodic boundaries impose symmetry particularly with respect to the instabilities taking place at the crown's rim not often observed in the experiments.

The adopted mesh resolutions and the level of detail of the numerical methods are in the majority of the studies not sufficient to predict the crown's breakup process. On the other hand, breakup shapes that have been numerically observed are often not found in the experimental visualizations.

With regard to the reviewed numerical work, it can be concluded that the application of the VOF method is considered to be the suitable tool for the direct simulation of the drop impact dynamics. However, further provisions are necessary like an enhanced level of detail for the correct resolution of the disintegration process with simultaneously a reduction of the overall computational time and effort.

2.8 Modelling Spray Impact

In addition to the direct numerical simulation, a far more established approach for the prediction of the spray-film interaction is the integration of empirical or semi-empirical drop-film models in computational methods that employ the Eulerian-Lagrangian method. An empirical model is then defined as such to relate post impingement quantities to pre-impingement parameters of discrete drop or parcels impingements. Extensive reviews of available empirical drop impingement models were conducted by e.g. Cossali et al. (2005). The available empirical formulations are mainly sub-divided into models collecting empirical results and generating a reasonable data base of post impingement quantities or models derived from individual experiments. This section gives only a brief overview of drop impingement models.

Bai and Gosman (1995) presented in their publication a model considering stick, rebound, spread and splash as impingement regimes. The thresholds between the impact regimes were taken from the work of Jayaratne and Mason (1964) and Stow and Hadfield (1981). The characteristics of the secondary drops and the ejected mass are obtained randomly independent on the input parameters. The model treats the impact onto a liquid surface as one onto a very rough wall without taking the liquid film height into account.

A model for spray impingement onto walls and thin liquid films ($H^* = 0.17$) including the effect of impact frequency was developed by Stanton and Rutland (1998). The model classifies the impingement regimes similar to that of Bai and Gosman (1995), namely stick, bounce, spread and splash. The transition to the splashing regime, and the secondary drop characteristics after breakup are taken from Mundo et al. (1995) and Yarin and Weiss (1995). A Weibull distribution was applied to fit the size and velocity distribution. The mean value and the deviation parameter to obtain the full Weibull distribution were depending only on the impacting drop Weber number.

Samenfink (1997) and Samenfink et al. (1999) developed an empirical spray impingement model for drop interaction with shear driven liquid films ($0 < H^* < 3$) from own measurements conducted with water. Two impingement regimes are considered, spread and splash, and a splashing threshold was obtained depending on the Reynolds and Laplace numbers. The secondary drop size and velocity (component) distributions were fitted using log-normal distribution functions.

The geometric mean quantity and variance of the distributions, the deposited mass fraction and the composition ratio were related to the impacting Reynolds and Laplace numbers, impingement angle and liquid film height.

A drop impingement model for drop splashing interaction with wetted surfaces ($H^* = 0.03$), and derived from experimental data of Mundo et al. (1995) was implemented by Mundo et al. (1997). A splashing threshold was proposed using the non-dimensional momentum, K. This quantity was then used to formulate correlations for the deposited mass fraction, the number, the mean size and velocity as well as the related variances of the secondary droplets.

In Cossali et al. (1999) and Cossali et al. (2005) another single drop impingement model is denoted called the Marengo-Tropea model. The model is derived from own phase Doppler anemometer measurements conducted during water drop impact on a moving and uniform film $H^* < 2$ and predicts the size and velocity distribution as well as the number and deposited mass fraction of the secondary drops.

With the help of the physical understanding derived from their analysis of the transverse instability driving the crown's breakup and the phase Doppler measurements of a sparse spray impact on a spherical target, Roisman et al. (2006) further developed some scaling laws and a semi-empirical model for the average secondary drop diameter, the secondary drop velocity and the volume flux density. The secondary droplet characteristics were correlated using only the impinging Reynolds number and for two parameter ranges, namely for $Re < 500$ and $Re > 500$. On the other hand, the correlations for the volume flux density additionally contained the K number. With increasing Reynolds number, a clear decrease of the secondary drop sizes was predicted while the volume flux increased with increasing K number

O'Rourke and Amsden (2000) improved an existing thin film model with a new spray-wall interaction model for application to port-injection engines. They developed a splash criterion depending on the film thickness, the drop velocity normal to the wall, the drop diameter and the fluid properties, respectively. A detailed parametrization of the experimental data of Yarin and Weiss (1995) and Mundo et al. (1995) describing secondary drop characteristics and deposited mass fraction is also proposed, however, using a new splashing input parameter E. Nearly at the same time Han et al. (2000) suggested a similar model using the same experimental sources as O'Rourke and Amsden (2000). The only difference occurs in a new and more accurate parametrization procedure of the secondary drop distributions and deposited mass fraction of the source data.

Limitations of the Reviewed Spray-Wall/Film Interaction Models

By analysing the available drop-wall and drop-film interaction models in literature, particularly with the objective of the application to bearing chamber CFD methods, several conclusions can be drawn. All of the investigations used an inadequate number of test fluids for their empirical models. This in turn caused a very narrow range of validity of the correlations. Particularly for the fluid properties found in bearing chambers none of the drop-film interaction models proved to be valid. Drop-film interaction models involving the impingement regimes of drop impacts

onto thicker liquid films are moreover very scarce in literature. Cossali et al. (2005) stated in
their review that most of the developed models derived from single drop impingement data were
conducted with millimeter size drops thereby not for drop sizes usually occurring in practical
situations. As already depicted in section 2.6.3, the measurement of the ejected mass fraction
during drop impact onto liquid surfaces is often inaccurate and results therefore also for drop-film
interaction models in non-physical outcomes. In this regard, Cossali et al. (2005) compared also
most of the available models. Figs 2.14a and 2.14b depict an exemplary comparison of the most
relevant drop impact models for an impingement of an engine oil droplet onto a wall and thin film,
respectively. The impact velocity is varied here between 5 m/s and 30 m/s in order to analyse the
predictions of the respective models. The comparison shows evidently that a large scatter of the
resulting ejected mass fraction is predicted by the models. Here, the models of Samenfink (1997)
and Stanton and Rutland (1996) appeared to be the most consistent but denoted a very narrow
range of validity. The widely distributed model of Bai and Gosman (1995) is not included into the
comparison since their model assumes a randomly changing ejected mass fraction independent
on the impacting parameter and only bounded by a maximum and a minimum value. All the
other authors noticed a large scattering of their predicted results for equivalent impact conditions
inside their validity range and serious inappropriate predictions outside of their range of validity.

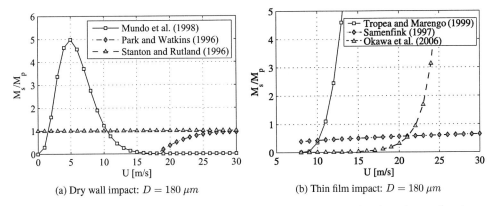

(a) Dry wall impact: $D = 180\ \mu m$

(b) Thin film impact: $D = 180\ \mu m$

Fig. 2.14: Drop-wall and drop-film interaction model comparison via the ejected mass fraction
for a typical bearing chamber oil drop impact scenario

Hence, in the light of the discussion, there is an absolute need for a drop-film interaction model
applicable in future bearing chamber design tools. The model should not only meet the parameter
range of bearing chambers but also ensure accurate predictions of the ejected mass fraction and
the secondary drop characteristics.

2.9   Concretization of Objectives and Aim of Investigation

The main objectives of the present thesis are, first, the development of a detailed physical understanding of the oil droplet impact dynamics. With the help of an improved understanding, then, a
drop-film interaction model valid for the range of impingement conditions found in the bearing
chambers is aimed at. Based on the previous literature of

- the range of drop-to-film impingement conditions found in aero-engine bearing chambers,
- the single drop impact dynamics,
- and the spray impingement modelling derived from single drop impact data,

several points and gaps have been identified in the light of application to bearing chamber
impingement conditions. With the help of the range of impingement conditions found in bearing
chambers and the characterization of the impingement regimes in the beginning of this chapter, it
was clearly shown that most of the droplets are impinging onto liquid wall films ranging from thin
films to shallow and deep pools. Moreover, a comparison of the impingement conditions with
the transitional regime threshold indicated that only the spreading and splashing of droplets with
liquid wall films is of frequent occurrence in bearing chambers. This fact demonstrates the need for
a comprehensive understanding of the splashing regime in particular, which requires knowledge
about the mass and momentum of the ejected secondary droplets. However, by reviewing the
literature dedicated to the single drop impact dynamics, it was quickly observed that only a very
narrow range of parameters in terms of the adopted test fluids, the impinging drop sizes, and the
target liquid wall film heights have been investigated.

With respect to the modelling of the influence of the wall film height on drop impingements, it has
clearly been shown that the role of the film thickness is not simply described by using the relative
film thickness H^*, but needs additional information about the energy of the impinging droplet.
For instance, a droplet impact event with a given liquid film height and different kinetic energies
will be affected by the wall only when the penetrating cavity reaches the bottom wall during the
impingement. The traditional classification using solely the relative film height H^* does not take
into account the effect of the kinetic energy and the respective penetration depth. Despite the
success of Bisighini (2010) for a better classification of the target liquid surface characteristics
using millimetre sized drop impingements and the cavity evolution, there have not been, to date,
any quantitative investigations applying this approach to the influence of the wall film height on
secondary droplet generation. Although the modelling of the cavity penetration dynamics needed
to characterize the influence of the liquid film height is very rich in the scientific literature, only
large millimetre sized droplets have been used to derive the theoretical descriptions, which are far
from the drop sizes identified in bearing chambers. Hence, there has been a lack of knowledge,
to be addressed here, as to the question of whether the application of millimetre sized drop
impingements can be used as a substitute for submillimetre sized drop impingement, in particular
with respect to the role of the Froude number. It was also seen that there has been very little
knowledge of the physical processes involving the role of the impingement angle in single drop
impact dynamics.

The experimental measurement techniques that have been adopted (imaging or laser optics) for the characterization of the secondary droplets and ejected mass fraction contain several inaccuracies, which are clearly confirmed by the large scattering of the correlation results for the ejected mass fraction of the most known drop-film interaction models. The direct numerical simulation of the splashing impingement was found to be a remedy for these experimental limitations. However, the numerical model setup, together with the level of resolution adopted, appeared to be often inadequate. In cases when the numerical simulations delivered very good results in terms of the splashing products, computational times proved to be impractical for the large parameter studies required for correlation development. In order to complement the gaps identified in the literature survey of the present thesis, a numerical methodology is proposed in Chapter 3 for the direct simulation of single drop impingement dynamics. This numerical methodology is capable of simulating the complete disintegration process in the splashing regime with less computational effort than existing approaches. It provides all relevant flow quantities characterizing the impingement process, enabling the direct evaluation of the products of splashing. The method includes a two-dimensional axisymmetric and three-dimensional Volume-Of-Fluid method coupled with an efficient Adaptive Mesh Refinement technique (AMR). The local mesh refinement technique enables refining the region where large resolution is required while keeping the insignificant regions of the flow field coarse. Since it has been shown that a large part of the impingement evolution, such as the lamella and cavity expansion, is two-dimensional, it is sufficient to apply a two-dimensional axisymmetric method for its evaluation. The three-dimensional breakup of the crown and the secondary drop generation is a 3D problem. Hence, the 3DVOF-AMR method will be adopted. In addition to the numerical methodology, due to the lack of accurate validation data, particularly for the crown dynamics, secondary drop characteristics, and ejected mass fraction following a single drop impact, an experimental data base was also developed in the present investigation with the help of an advanced imaging technique. The experimental method together with the validation of the numerical method is presented in Chapter 4. After the validation of the numerical method, the effect of different non-dimensional parameters on the secondary drop characteristics during single drop impingement with and without the influence of the bottom wall will be presented for the range of parameter found in bearing chambers (in Chapters 5 and 6). With the help of the simulated and measured data base, suitable correlations are derived in Chapter 7, building a base for the targeted drop-film interaction model for application to the numerical simulation of bearing chambers. The drop-film interaction model will comprise the following points:

- classification of the influence of the bottom wall on the impingement outcome,

- prediction of the impingement regime using a transitional regime threshold,

- prediction of the secondary droplet characteristics, namely the distribution of the number, size, and velocity (components) of the secondary drops and the deposited mass fraction.

3 Numerical Method and CFD Modelling of Single Drop Impacts

In the last chapter, the limitations of the experimental investigations for measuring the secondary droplet characteristics and the ejected mass fraction during single drop impact in particular have been clearly evidenced. The direct numerical simulation of the crown's disintegration following single drop impingement was proposed as the future remedy and opened new horizons for the evaluation of the splashing products. However, shortcomings in the numerical resolution, inaccurate solution procedures and the large computational times make the direct simulation of the splashing products still an insurmountable step.

This chapter proposes therefore an efficient numerical simulation technique based on the Volume-Of-Fluid method and extended with an Adaptive Mesh Refinement technique. This technique enables to concentrate the level of computational detail in areas, where it is needed, by reducing the computational effort at the same time. With this method in hand, a detailed simulation of the cavity evolution, the lamella expansion and the three-dimensional crown breakup dynamics will be achieved. The chapter starts with the mathematical formulation of the equations governing the drop impact dynamics. Then, an adequate numerical solution procedure for the differential equations is described. At the end of this chapter, the two-dimensional axisymmetric and the three-dimensional domains are discussed together with the related boundary conditions.

3.1 Mathematical Method - Volume-Of-Fluid Method

The dynamics of single droplet impingement onto liquid surfaces can be considered as a multiphase problem, occurring with two or more immiscible fluids with a free surface. One approach that is usually known and applied in literature is given with the *homogeneous Eulerian* Volume-Of-Fluid method of Hirt and Nichols (1981). The system of partial differential equations in 3.1-3.2 is the mathematical formulation for the description of two-phase drop impact dynamics in the present investigation. The laminar and incompressible (for each phase) partial differential equations are defined by the continuity equation,

$$\nabla \cdot \vec{v} = 0, \tag{3.1}$$

and the momentum transport equation,

$$\frac{\partial(\rho\vec{v})}{\partial t} + \nabla \cdot (\rho\vec{v} \otimes \vec{v}) = -\nabla p + \rho\vec{g} + \nabla \cdot \tau + \vec{f}_\sigma. \tag{3.2}$$

Body forces due to the compression force ∇p and gravity $\rho\vec{g}$ are implicitly represented by the first two terms on the r.h.s of equation 3.2. Friction and surface tension forces are expressed by the divergence of the stress tensor and the term $\vec{f}_\sigma$, respectively.

A characteristic feature of the VOF method is the introduction of a phase indicator function α in the single phase Euler equations. The transport equation for the indicator function α is

$$\frac{\partial \alpha}{\partial t} + \nabla \cdot (\vec{v}\alpha) = 0, \tag{3.3}$$

The indicator function may have different values during the simulation with

- $\alpha = 0$, for cells belonging to the gaseous phase,
- $\alpha = 1$, for cells points belonging to the liquid phase,
- $0 < \alpha < 1$, for cells containing both phases.

It may hence be interpreted as the volume or phase fraction of one phase as

$$\alpha = \frac{\int_V \alpha_l}{\int_V (\alpha_g + \alpha_l)}, \tag{3.4}$$

where the subscripts l and g denote the liquid and gas and $\alpha_g + \alpha_l = 1$. The discontinuity of fluid properties across the interface are modelled similar to the indicator functions as continuous in time and space in the governing equations and defined with

$$\rho = \alpha \cdot \rho_l + (1 - \alpha) \cdot \rho_g, \tag{3.5}$$

and

$$\mu = \alpha \cdot \mu_l + (1 - \alpha) \cdot \mu_g. \tag{3.6}$$

Numerical simulations of free surface flows using the VOF method are, hence, approximating the discontinuity across the interface by a continuous approximation solving the partial differential equation of the indicator function. This causes typically not a sharp interface but a more or less smeared resolution over several cells due to discretization errors (numerical diffusion). The selection of an appropriate spatial resolution and discretization of the convective term within the transport equation of the indicator function is therefore crucial in particular for two-phase flows with high density ratios. An inadequate number of discretization points near the discontinuous interface may lead to significant errors in the evaluation of the surface curvature and thereby the surface tension forces driving the dynamics of drop impact. The surface tension force $\vec{f_\sigma}$ in the momentum equation in 3.2 is accounted for by using the continuum surface force (CSF) approach of Brackbill et al. (1992). This concept treats the discontinuous surface tension force acting usually as a surface force at the interface as a volumetric force located in those cells of the mesh containing the interface region. The volumetric force can hence be added in the momentum equation as an ordinary source term

$$\vec{f_\sigma} = \sigma \kappa \vec{n}, \tag{3.7}$$

where σ represents the surface tension coefficient, κ the surface curvature and $\vec{n}$ the interface normal vector pointing from the heavy fluid to the lighter fluid. The normal vector is calculated from the gradient of a smoothed volume fraction with

$$\vec{n} = \frac{\vec{\nabla}\alpha}{|\vec{\nabla}\alpha|}. \tag{3.8}$$

The curvature κ is given by the divergence of the normal vector

$$\kappa = \vec{\nabla} \cdot \vec{n}. \tag{3.9}$$

Wall adhesion effects are not considered in the present study since the focus was put on drop impingement onto thick liquid surfaces. For these impact scenarios, the effect of wall adhesion plays a minor role.

The evaluation of the friction force in equation 3.2 given by the divergence of the stress tensor, simplifies to

$$\vec{\nabla} \cdot \tau = \vec{\nabla} \cdot \left(\mu \left[\vec{\nabla} \otimes \vec{v} + \left(\vec{\nabla} \otimes \vec{v} \right)^T \right] \right) = \vec{\nabla} \cdot \left(\mu \vec{\nabla} \otimes \vec{v} \right) + \left(\vec{\nabla} \otimes \vec{v} \right) \cdot \vec{\nabla} \mu, \tag{3.10}$$

since both fluids are considered to be of Newtonian nature and incompressible (see Ferziger and Peric (2008)).

3.2 Numerical Solution Procedure

3.2.1 Discretization Procedure

The transport equations in section 3.1 are non-linear, second order partial differential equations. The variation of the flow quantities in space and time is therefore continuous. To date, a solution of such equations can only be found numerically by adopting a discretization procedure. In the present thesis a finite-volume discretization method is adopted to transform the partial differential equations into discrete forms, integrated in space over each discrete volume.

3.2.1.1 Discretization of the Spatial Derivatives

The discretization of the spatial derivatives is conducted in the present investigation on a computational domain that is subdivided into a number of uniform, structured hexahedral volumes with computational points placed at cell-centroids. The transport equations are valid for each cell. Each cell contains its computational points at its cell-centroid stored and a finite number of flat faces f. The normal vectors n_i of the cell faces and the indices of the neighbour cells are also depicted. During the computation of a fluid dynamic problem, the transport equations are solved by integration over the volume:

$$\int_{Vol} \frac{\partial \rho \phi}{\partial t} dVol + \oint \rho \phi \vec{v} \cdot d\vec{A} = \oint \Gamma_\phi \cdot d\vec{A} + \int_{Vol} S_\phi dVol. \tag{3.11}$$

$\phi(\vec{r},t)$ is a general flow quantity, Γ_ϕ is an appropriate diffusion coefficient and S_ϕ is the local source term. With the help of the Gauss's theorem (Ferziger and Peric (2008)), the convective and diffusive terms can be converted into surface integrals over the surfaces, which can be simplified under the assumption of linearity as a sum over six surfaces of the finite volume.

$$\frac{\partial \rho \phi}{\partial t} \cdot Vol + \sum_f^{N_f} \rho_f \phi_f \vec{v_f} \cdot \vec{A_f} = \sum_f^{N_f} \Gamma_\phi \cdot \nabla \phi_f \cdot \vec{A_f} + S_\phi \cdot Vol \tag{3.12}$$

N_f represents the number of cells and ϕ_f represents the value of the flow quantity at the face f. $\rho_f \vec{v} \vec{A}_f$ indicates the mass-flow through the cell face f and Vol denotes the volume of the cell. At this stage it is possible to determine quantities for the face $\vec{A}_f$ of the cell. However, the surface integrals of the flow quantity ϕ are calculated only in the cell-centroid of the corresponding finite volume.

Cell-face interpolation

In order to compute the values of the convective and the diffusive terms in equation 3.11 at the centers of the cell-faces, as required after application of the Gauss's theorem, additional interpolation is needed. The interpolation procedure delivers the flow quantities ϕ_f at the cell-faces from the calculated cell centroid values ϕ of the neighbour cells. In the present investigation the interpolation of the convective and diffusive terms in the continuity and momentum transport equation is carried out using a second order upwind scheme (see Ferziger and Peric (2008)). When second order accuracy is required or desired, the flow quantities ϕ_f are computed using a multidimensional linear reconstruction scheme developed by Barth and Jespersen (1989). In this approach, higher-order accuracy is achieved at cell faces through a Taylor series expansion of the cell-centroid solution. The face value ϕ_f is hence computed with

$$\phi_f = \phi + \nabla\phi \cdot \vec{r}, \tag{3.13}$$

where ϕ and $\nabla\phi$ are the cell-centered value and its gradient in the upstream cell, and $\vec{r}$ is the displacement vector from the upstream cell-centroid to the face centroid. The advantage of a second order upwind scheme compared to the usually used standard central differencing scheme lies in the boundedness of the solution provided by the upstream-biased numerical solution of the convection term (see Ferziger and Peric (2008)). However, by using only low order interpolation schemes numerical diffusion may become significant.

Source term

The discretization of the source term in equation 3.12 is obtained by assuming the source term value to be constant within the entire finite volume of the cell. Thus the integral in equation 3.12 can be estimated with the source term and the volume of the cell with

$$\int_{Vol} S_\phi dVol \approx S_\phi \cdot Vol. \tag{3.14}$$

Interpolation near gas-liquid interface

The interpolation and reconstruction of the gas-liquid interface is carried out in two stages using a Donor-Acceptor Scheme (DAS) and a second order reconstruction scheme based on the slope limiter (see Hirt and Nichols (1981) and Ferziger and Peric (2008)). In the commercial CFD Software Ansys Inc. this integrated scheme is usually referred to as the *compressive scheme*.

The DAS calculates the amount of fluid advected through the faces by identifying one cell as a donor of an amount of fluid of one phase and another neighbouring cell as the acceptor of an equivalent amount of fluid. With this method face fluxes are computed based on the interface orientation, hence, either horizontal or vertical depending on the volume fraction gradient $\nabla \alpha$ between two neighbouring cells. The interface orientation after the calculation of the phase fraction with the DAS is mostly discontinuous and not suitable for a correct computation of the surface curvatures. An additional second order reconstruction scheme is therefore applied that uses the previous phase fraction values and reconstructs the interface with a slope limiter avoiding over- and undershoots of the phase fraction values. The interpolation method can be defined with the face value of the phase fraction as follows

$$\phi_f = \phi_{DAS} + b\nabla\phi_{DAS}, \tag{3.15}$$

where ϕ_{DAS} and $\nabla\phi_{DAS}$ are the donor cell phase fraction value and its gradient. b represents the slope limiter value. In the present investigation all the simulations were carried out with a slope limiter value of $b = 2$ corresponding to higher order interpolation. The reason why this method is favoured over the usually known geometric reconstruction scheme PLIC (see Rider and Kothe (1998)), or the compressive interface capturing schemes for arbitrary meshes CICSAM (see Ubbink and Issa (1999)) because better coupling with a dynamic adaptive mesh refinement (AMR) technique to be presented in the next section in 3.2.3 is achieved. As will be shown later in this chapter, the gradient adaptive mesh refinement technique relies in particular on a minimum amount of numerical diffusion in order to obtain a bend of refined mesh resolution near the free surface. The increased resolution near the interface using the AMR method combined with the higher order compression scheme allows then a very accurate computation of the surface curvature and, as a consequence, of the surface tension forces driving the drop impingement dynamics.

3.2.1.2 Discretization of the Temporal Derivatives

The remaining temporal derivatives of equation 3.11 need to be discretized in time. In the present investigation the integration of the terms in equation 3.11 over time is accomplished by the Euler implicit bounded differencing scheme available in the CFD software Ansys Inc. This method is second order accurate in time and guarantees boundedness of the solution. The spatial integration of the transient flow quantity can be approximated in the same way as in Equation 3.14, assuming that the flow quantity is not changing within the cell during a time step.

$$\int_{Vol} \frac{\partial\left(\rho\phi\right)}{\partial t}dVol \approx \frac{\partial\left(\rho\phi\right)}{\partial t} \cdot Vol = F(\phi), \tag{3.16}$$

where the function $F(\phi)$ incorporates any term of the transport equations discretized in space. The integration of the transient terms is straightforward and calculated as

$$\frac{\partial\left(\rho\phi\right)}{\partial t} \approx \frac{3\left(\phi\rho\right)^{n+1} - 4\left(\rho\phi\right)^{n} + \left(\rho\phi\right)^{n-1}}{2\Delta t}, \tag{3.17}$$

where $n + 1$ is the value at the next time level, $t + \Delta t$, n is the value at the current time level, t, $n - 1$ is the value at the previous time level, $t - \Delta t$.

Bounded Second Order Implicit Time Integration

The time level values of ϕ used for computing the function F is calculated at a future time level $n + 1/2$ with

$$F(\phi_{n+1/2}) = \frac{\phi_{n+1/2} - \phi_{n-1/2}}{\Delta t}, \tag{3.18}$$

where

$$\phi_{n+1/2} = \phi_n + \frac{1}{2}\beta_{n+1/2}(\phi_n - \phi_{n-1}), \tag{3.19}$$

and

$$\phi_{n-1/2} = \phi_{n-1} + \frac{1}{2}\beta_{n-1/2}(\phi_{n-1} - \phi_{n-2}). \tag{3.20}$$

$\beta_{n+1/2}$ and $\beta_{n-1/2}$ are bounding factors for each variable at the $n + 1/2$ and $n - 1/2$ level. In the present investigation all the calculated transport equations are bounded with a lower bound of 0 and an upper bound of 1.

Time step selection

A common way to select an appropriate time step in CFD simulations is given by applying the Courant-Friedrichs-Lewy (CFL) condition

$$\Delta t < \frac{l_{cell}}{|\vec{v}|}, \tag{3.21}$$

with the time step Δt, the edge-length of the smallest cell volume l_{cell} and $|\vec{v}|$ the averaged velocity scale. Many authors showed in VOF-based methods, however, that convergence and stability of the solution of the phase fraction transport equation could only be obtained for maximum Courant number much below unity. One reason for such low Courant number is related to the application of a CSF model as a continuous volume force in the momentum equation. Numerical errors during the solution of the momentum equation in the transitional area of the interface cause so-called spurious or parasitic currents (see Harvie et al. (2006) and Galusinski and Vigneaux (2008)). Two-phase flows with high density ratios and surface tension effects (capillary driven flows) in particular proved to generate artificial velocities with non-physical instabilities at the interface. Galusinski and Vigneaux (2008) proposed in their publication a new stability condition for the time step selection in capillary driven two-phase flows with low and medium Reynolds numbers to reduce the magnitude of the these spurious currents with

$$\Delta t = \frac{\mu \cdot l_{cell}}{\sigma}. \tag{3.22}$$

As will be shown later, using this capillary time-step the resulting Courant Numbers were at least one order of magnitude smaller than 1 ($CFL \approx 0.06$).

3.2.2 Pressure-Velocity Coupling

All the fluids in the present investigation are of Newtonian nature and incompressible. This principally simplifies the computational solution procedure of the transport equations since no explicit equation for the solution of the pressure is required. In order to avoid a de-coupling of the pressure from velocity, however, a discretized pressure equation is derived from the semi-discretized momentum equation using the continuity restriction in equation 3.1. In the present investigation coupling between pressure and velocity is performed through the Pressure Implicit with Splitting of Operators (PISO) algorithm for transient flows of Issa (1986). This coupling method was already successfully adopted recently among others by Berberović et al. (2009) for drop impacts onto liquid layers of finite thickness and proved also in the present study to be accurate.

3.2.3 Adaptive Mesh Refinement

The development of a numerical method capable to resolve the splashing dynamics of drop impacts onto liquid surfaces is still very challenging and scarce in the literature. The crown's breakup, for instance, takes place with temporal and spatial length scales that are of several orders of magnitude smaller than the duration of the complete impingement evolution. This fact implies the application of very large computational mesh resolutions, with small time steps and simultaneously long impingement duration times. Authors as Rieber and Frohn (1999), among others, have indeed been able to successfully calculate the crown's disintegration with VOF-based methods but on computational domains having up to several hundreds of million cells (for just one quarter of the impingement evolution). Such a large number of computational cells is up to date still impractical for simulation of various impingement events needed for the development of correlations for drop-film interaction.

In the present investigation the usage of an Adaptive Mesh Refinement (AMR) technique available in the commercial software package *Ansys Inc Fluent* 14.5 enabled to significantly reduce the computational effort of the simulations by increasing the mesh resolution transiently at locations were large gradients of the flow quantities occurred while keeping the other parts coarsened. Similar numerical procedures have successfully been applied to similar problems by Nikolopoulos et al. (2007), Popinet (2009), Fuster et al. (2009), Fest-Santini et al. (2012), and Brambilla and Guardone (2013). However, the present methodology, as a difference to the previous investigations, combines a rather coarse interface interpolation scheme (DAS) together with the gradient adaptation technique described later in this chapter. Such a combination was used on purpose by the author for two reasons. On the one hand it was used to generate a broad band of disturbances acting throughout the flow field (where $|\nabla \alpha|$ is zero). These disturbances are needed to activate the peak wave length of the instability driving the breakup of the crown's rim. The selection of very accurate interface reconstruction schemes such as the CICSAM scheme in Nikolopoulos et al. (2007) or the Geometric Reconstruction scheme in Gomaa et al. (2009) have not been successful with respect to the calculation of the crown's disintegration process usually observed in experiments.

On the other hand, a minimum amount of numerical diffusion in the areas $0 < \alpha < 1$ enabled an improved control of the gradient mesh adaptation method. The AMR technique used in *Ansys Inc Fluent* 14.5 is based on a hanging node adaptation procedure. Quadrilateral (2D)/ hexahedral (3D) cells are marked and refined by splitting the cell into 4 or 8 cells generating so-called hanging nodes on edges and faces that are not vertices of all the cells sharing those edges or cells. Accordingly, coarsening can also take place by removing the hanging nodes and re-unifying the *child* cells to *mother* cells. For each coarsening or refining step a new mesh hierarchy is generated and saved in additional memory. The number of splitting/re-unification procedures and hanging node production/elimination can be adjusted by setting the level of refinement for the specific problem. Several criteria can be used in *Ansys Fluent* for the pre-marking of cells that need to be refined for each time step, namely the gradient, the curvature and the iso-value approach. In the present study only the gradient approach was applied. It will be discussed in more detail in next paragraph. The gradient approach used in the present study takes the Euclidean norm of the phase fraction gradient $|\nabla\alpha|$ and multiplies it with a characteristic length (see Daunenhofer and Baron (1985))

$$|\nabla\alpha|A_{cell}^{\frac{r}{2}} < \epsilon, \tag{3.23}$$

where ϵ is the threshold parameter , A_{cell} is the cell area and r is the gradient volume weight. The threshold parameter is selected as a normalized value despite a standard absolute value. The normalization is carried out as

$$\frac{|\epsilon|}{max|\epsilon|}. \tag{3.24}$$

With this criterion it is possible to generate a band of constant high resolution of cells in the vicinity of the interface defined by the volume fraction α. Depending on the numerical error of the second order spatial and temporal discretization of the phase fraction transport equation, the threshold parameter ϵ, the size of the band of high resolution ($|\nabla\alpha|$ is not zero) may be adjusted according to the requirements of the fluid dynamic problem.

3.3   Computational Domain and Initial/Boundary Conditions

Numerical simulations of single drop impingement onto liquid surfaces are aimed with particular focus on the three-dimensional secondary drop generation. However, in order to characterise the effect of the wall (film-depth) on the secondary drop generation, a detailed understanding of the interaction between the cavity and crown evolution together with the maximum cavity depth for deep pool impacts is crucial (see Chapter 2). For this objective, the review of numerical investigations has shown that two-dimensional axisymmetric numerical methods are enough capable to resolve the cavity and crown expansion dynamics during single drop impingement.

Hence, the description of the numerical setups used in this thesis is divided into two parts. It reports first about the two-dimensional axisymmetric numerical setup applied for the numerical simulation of the crown and cavity expansion. Then, the computational domain and boundary conditions of the three-dimensional methodology is conducted providing all the details for the simulation of the secondary drop generation due to the crown's disintegration.

3.3.1   2D-VOF-AMR - Numerical Model for Crown and Cavity Expansion

As depicted in Fig. 3.1, the two-dimensional computational domain used for the simulations consists of an axisymmetric slice with dimensions in the direction perpendicular to the liquid surface of $Y = 24 \cdot D$ and parallel to the liquid surface of $X = 37 \cdot D$. The choice of an axisymmetric formulation of the governing equations was selected according to several experimental observations in the literature related to the cavity and lamella rim evolution during single drop impact.

The grid was divided in two different zones in horizontal direction at approx. $12 \cdot D$ having an interface (see Figure 3.1b). The distance between the uniform interface and the impingement

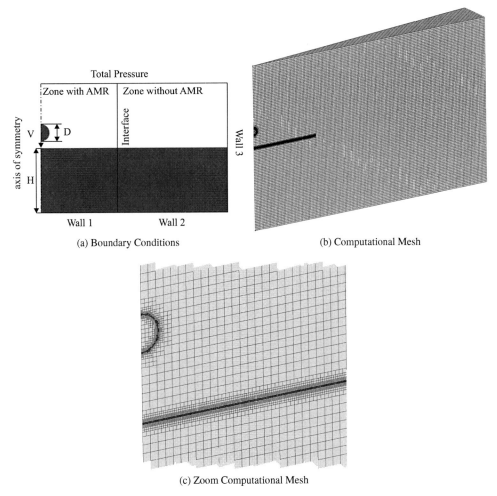

(a) Boundary Conditions

(b) Computational Mesh

(c) Zoom Computational Mesh

Fig. 3.1: Boundary Conditions and Computational Mesh for the investigation of the cavity and lamella rim evolution

Parameter	No. of cells (AMR)	No. of cells (full)	No. of processors	Comp. time
We ↑ & Fr ↓	≈ 26k	≈ 900k	4	≈ 150h
We ↑ & Fr ↑	≈ 26k	≈ 900k	4	≈ 240h

Table 3.1: Example of the required computational resources for two different impingement conditions

region was $12 \cdot D$. An interface at this position enables to define two computational domains with two different grid resolutions.

Within the zone containing the droplet (internal zone), an AMR technique is adopted leading to a maximum level of refinement of 4 near the free-surface. In the external zone, the same (coarse and uniform) resolution throughout the duration of the complete simulation (see Fig. 3.1b) is used. The transition from a very dense to a coarse mesh resolution causes user-induced numerical diffusion at the interface. For this case, the numerical diffusion has a positive impact onto the simulation accuracy since it avoids the reflection of capillary waves at the side boundaries. This in turn removes the effect of the side boundary conditions on the impingement evolution.

An initial base grid of $17.3k$ uniform quadrilateral cells was used for all simulations. This mesh was dynamically refined every $8th$ time-step near the free surface using the gradient adaptive method presented in the last chapter. When the normalized gradients of the air volume fraction near the free surface resulted in values smaller than $\nabla \alpha < 0.05$, a coarsening procedure took place where the previously generated hanging nodes within the refined cells were reunified. Accordingly, for $\nabla \alpha > 0.15$, a refinement of the mesh near the interface was conducted. The level of refinement necessary for an accurate resolution of the temporal and spatial evolution of the cavity and lamella rim was based on both experience of previous studies of Berberović et al. (2009) and a separate mesh independence study. For the latter, numerical simulations with different levels of refinement (2,3 and 4) were conducted and compared to available experimental results. The results revealed that the smallest cell edge length required near the free surface for resolving the dynamics of interest was approximately $(D/90)$. For the latter resolution, the total number of cells resulted in ≈ $26k$, compared to more than ≈ $900k$ with a corresponding uniform resolution grid. Fig. 3.1 depicts the mesh with the smallest resolution of $(D/90)$.

The liquid film and the impinging droplet were set at the beginning of the simulation with the help of the volume fraction distribution. The droplet was located near the free surface at approx. $1D$ distance in y-direction from the horizontal free surface. At this position, any additional acceleration or deceleration due to gravity or drag is negligible, but there is still enough computational time available to calculate an accurate drop curvature and pressure distribution before impact. The impingement velocity of the droplet was accordingly set equal to the experimental test conditions.

The computational domain is bounded at the bottom and the side by walls with no-slip conditions and total pressure inlet conditions at the top. The latter allows an adjustment of the static pressure at the boundary based on the calculated velocity field. Table 3.1 indicates the computational times needed for one impingement event for a low and a large Froude number impingement. For low Froude numbers, the gravitational or centrifugal acceleration is able to pull the free surface quicker towards its original position. The strong retraction forces coming from the potential energy of liquid film cause a faster retraction of the cavity and the lamella. This leads to a shorter computational time compared to large Froude number impingements.

3.3.2   3D-VOF-AMR - Numerical Model for Splashing

The disintegration of the crown's rim during single drop impingement onto liquid surfaces is a three-dimensional phenomenon that requires a 3D-VOF-AMR method.

For this purpose, a three-dimensional domain is generated in two distinct versions, one for the normal and one for the oblique impingement onto liquid surfaces. The computational domain for the simulation of the normal impact is a rectangular element with a quadrilateral section of $X = 9D$ and $Z = 9D$ and a height of $Y = 15D$ as depicted in Fig. 3.2a. The oblique impingement setup, on the other hand, must allow an enhanced evolution of the liquid film and the lamella rim in the direction of the velocity component parallel to the liquid surface. This is achieved by extending the domain dimensions to $13D$ in X-direction.

Two zones are assigned to the computational domain in both cases. The zones are separated by mesh interfaces (see Fig. 3.2a and 3.2b). The interior zones of both domains have equivalent dimensions with $X = 7D$, $Z = 7D$ and a height of $Y = 15D$. The splitting into two zones is needed to ensure that adaptive mesh refinement is conducted only in the interior zone where the splashing impingement takes place while maintaining the outer region coarsened throughout the impingement evolution. This provision ensures a forced numerical diffusion at the interfaces between inner and outer zone. Thereby the small scale capillary waves are dissipated and not reflected at the boundaries which might affect the crown's breakup.

The base spatial discretization of the zones is accomplished using $200k$ uniform hexahedral elements for the normal impingement and $450k$ cells for the oblique impingement setup. The corresponding cell edge length of the base grid resulted in $D/3$ for the normal impingement and $D/6$ for oblique impingement setup.

The liquid film and the impinging drop were again set at the beginning of the simulation with the help of the volume fraction distribution. The droplet was prescribed according to the axisymmetric simulations at approx. $1D$ distance in y-direction from the surface with a normal or oblique absolute terminal velocity. The meshes were then refined with a band of constant high resolution at the liquid free surface before starting the impingement, specifically with a level of refinement of 5 for the normal and 4 for the oblique impingement. With these resolutions, the minimum cell edge length at the free surface is $D/96$ leading to a starting number of 1 million and 1.5 million elements, respectively. In the present investigation only such a high resolution near the free surface ensured a comprehensive calculation of the crown disintegration during single drop impingement. Studies of Nikolopoulos et al. (2007) and Gomaa et al. (2009) on the effect of mesh resolution on the crown's disintegration stated similar terminal mesh resolutions.

During the simulation of the impingement evolution, the grid was dynamically refined every $8th$ time-step using the gradient adaptation method with equivalent $coarsen$ and $refine$ gradient thresholds as for the axisymmetric simulations in the previous section. The maximum number of

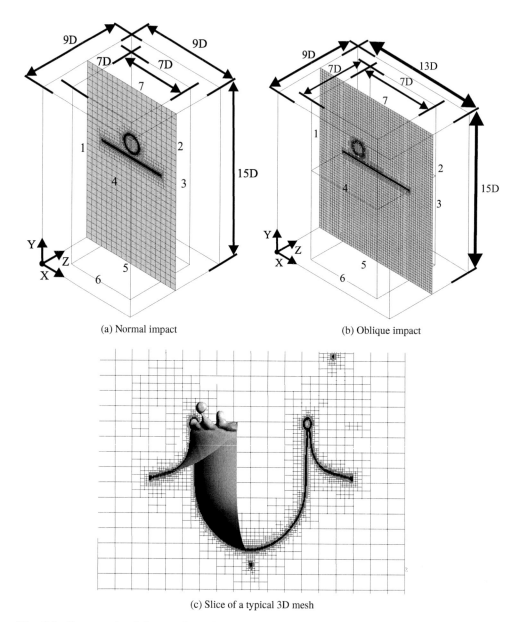

(a) Normal impact   (b) Oblique impact

(c) Slice of a typical 3D mesh

Fig. 3.2: Computational domain, boundary conditions and a typical 3D mesh for the normal and oblique impingement simulations

elements reached values around 5 million for the normal and 7 million for the oblique impingement, depending on the number and sizes of the secondary drops generated during crown breakup (see Table 3.2). Hence, the use of the present gradient VOF-AMR technique leads to a significant reduction of the computational effort. The significant increase in efficiency of the present

Impact	No. of cells (AMR)	No. of cells (full)	No. of processors	Comp. time
Normal	$\approx 5 \cdot 10^6$	$\approx 708 \cdot 10^6$	32	$\approx 450h$
Oblique	$\approx 7 \cdot 10^6$	$\approx 1.02 \cdot 10^9$	32	$\approx 672h$

Table 3.2: Example of the required computational resources for one impingement using both numerical setups

Number	Boundary condition	Number	Boundary condition
1	Symmetry	1	Tot. pressure inlet (open channel)
2	Symmetry	2	Symmetry
3	Symmetry	3	Stat. pressure outlet (open channel)
4	Symmetry	4	Symmetry
5	Wall	5	Wall
6	Interface	6	Interface
7	Tot. pressure	7	Tot. pressure

Table 3.3: Boundary conditions for the normal impact (left) and the oblique impact (right)

methodology becomes evident when taking into account that authors like Rieber and Frohn (1999) and Gomaa et al. (2009) used fully resolved meshes of 256^3 and 512^3 hexaedral elements for the direct numerical simulation of a splashing impingement of only a quarter section of the full problem. As stated in Table 3.2, full resolution meshes in the present thesis would result in $708 \cdot 10^6$ for the normal impingement and $1.02 \cdot 10^9$ for the oblique case. A typical time instance of the simulated crown rim is depicted in Fig. 3.2c using the iso-surface of the volume fraction at $\alpha = 0.5$. The mesh slice cutting the iso-surface shows the adapted mesh resolution with a very dense grid near the gas-liquid interfaces.

The computational domain of the normal impingement case was bounded at the bottom by no-slip conditions, at the side surfaces by symmetry conditions and a total pressure boundary at the top allowing for adjustment of the static pressure based on the calculated velocity field.

For the oblique impingement, symmetry boundary conditions are only defined at the surfaces at $Z = 0$ and $Z = 9D$ (see Fig. 3.2b). The side surfaces at $X = 0$ and $X = 13D$ are bounded using a total pressure and a zero pressure gradient boundary. The remaining boundaries are equivalent to the normal impingement case. Table 3.3 summarises the boundary conditions for both numerical setups. The indices of the different boundaries are illustrated in Fig. 3.2.

The literature review in chapter 2 has shown that most of the numerical investigations simulating the crown's disintegration during single drop impact and particularly the one using full resolutions required additional disturbances on the flow field in order to obtain physically reasonable results of the crown's breakup. Rieber and Frohn (1999), Fullana and Zaleski (1999) and Gomaa et al. (2009), for instance, imposed random disturbance to the initial velocities of the liquid film and the drop in each control volume. In the present 3D-VOF-AMR method, similar to the investigation of Nikolopoulos et al. (2007), no additional disturbances were initially imposed on the flow field. However, the use of a gradient adaptation method showed to provide a broad band of disturbances implicitly where no volume fraction gradient and zero free surface curvature occurred along the free surface. In this case, the variation of mesh resolution along the free surface caused the generation of small irregular and non-symmetric capillary waves propagating along the free surfaces. Occasionally, depending on the parameter of influence these small numerically induced capillary disturbances were magnified (when a resonance frequency existed) and/or damped in time.

3.4 Concluding Discussion

In summary, in this chapter, a brief description of the numerical methodology adopted for the two-dimensional and the three-dimensional simulations of single drop impingements onto liquid films have been presented. It included the mathematical formulation, the numerical solution procedures and the numerical simulation details, respectively. The Volume-Of-Fluid method is selected here (i) due to the good coupling capability with interface interpolation and curvature evaluation schemes, (ii) the simple connection with an Adaptive Mesh Refinement technique, (iii) due to the broad experience existing in the scientific literature. In particular, the coupling with an Adaptive Mesh Refinement technique will show that this numerical method is capable of simulating the complete three-dimensional disintegration of the crown's rim due to a very high resolution of the areas of interest. However, the computational effort will be reduced considerably compared to previous numerical investigations. Another interesting numerical feature is that no additional artificial disturbances need to be imposed on the flow field. The numerical method produces a broad band of disturbances without external intervention. These disturbances are absolutely needed to obtain plausible structures of the crown's disintegration shape.

Before discussing the effect of different parameters of influence on the cavity and crown dynamics during single drop impingement onto liquid surfaces, deriving correlations for secondary drop generation and build a drop-film interaction model, an experimental validation of the numerical methodology is inevitable. Unfortunately, there is very little data available in scientific literature for the crown disintegration shape, secondary drop characteristics and ejected mass fraction in particular. The cavity and crown expansion on the other hand appear to be well understood for large millimetre-sized drop impingements. Hence, in the light of discussion made, an experimental data base needs to be developed in this thesis that is used later for validation of the numerical method. The experimental validation including the experimental setup, the enhanced measurement technique and the comparison with numerical results is therefore presented next in this thesis.

4 Experimental Validation of the VOF-AMR Method

The review of the scientific literature focussing on the numerical simulation of drop impingements in Chapter 2 indicated that many authors have attempted to gain an improved understanding of the drop impact dynamics using the Volume-Of-Fluid method. However, only a few have been able to conduct a thorough verification and validation of their applied methodologies, and that mainly for the 2D cavity and crown dynamics rather than for the splashing products. This shortcoming is due to the very poor experimental data available for characterizing the secondary drop generation and ejected mass fraction during a single drop impingement.

Hence, this chapter proposes an experimental validation of the two methodologies adopted, namely the 3D-VOF-AMR for the simulation of the crown's disintegration and the 2D-VOF-AMR for the cavity penetration. The 3D-VOF-AMR method is validated using the present experimental results. The experimental validation is conducted on a test facility with well-defined boundary conditions and with an enhanced measurement technique. The latter could only be achieved by using millimetre sized single drop impingements onto deep liquid pools. The chapter starts with the description of the test facility and the adopted High-speed Shadowgraphy (HS) and Particle Tracking Velocimetry (PTV) techniques. The second part of the present validation chapter presents a thorough comparison of the numerical results using 2D-VOF-AMR with experimental data from the scientific literature.

4.1 Experimental Method

Following the goal to conduct an accurate experimental investigation targeting the measurement of

- the crown expansion,
- the secondary drop characteristics,
- and the ejected mass fraction,

during single drop impact onto liquid surfaces, the only experimental procedure for accessing the details of the process is a combination of digital high-speed imaging and image processing technique.

Figure 4.1 outlines the experimental setup, which consists of four main parts: (i) the drop generation, (ii) the target pool, (iii) the light source and (iv) the image acquisition system. A controlled single drop is generated by a syringe, falls down from a defined height and impacts onto a target deep pool.

A high intensity lamp ensures a strong continuous light that is diffused with an etched macrolon glass and provides a back-illumination of the impingement region. The images were recorded using a high-speed camera. Drop generation, target pool and the light source were mounted on a traversed platform allowing adjustment of the position of the impingement region, and control of the distance to the recording device. This capability is required, as will be discussed later, to record directly the depth-of-field of the optical setup and ensure that all the detected secondary drops are positioned within the depth-of-field. The components were fed, adjusted, controlled and synchronized manually with the help of a remote starting actuator for the recording device.

Drop Generation

A push-pull medical syringe was manually fed and controlled in order to dispense single drops. The drop is formed and grows at the tip of the needle and detaches when its gravitational forces exceed the capillary forces. The impacting drop size was varied by using various needles of diameter ranging between $0.6\ mm$ and $2.9\ mm$. Depending on the adopted fluids, however, there is a lower limit for the realizable drop size arising from the much faster reduction of gravitational forces with the size compared to the capillary forces. In this investigation, the smallest realizable drop size was $D = 2.5\ mm$. A maximum limit of the drop size is set by the sphericity. When a drop detaches from a needle its shape is not perfectly spherical. The competition between the viscous forces inside the droplet, the surface tension forces, the aerodynamic drag and the gravitational forces decides the level of drop deformation at its impact onto the liquid surface. As a result, the maximum drop diameter satisfying an acceptable degree of sphericity was found to be approximately $D = 4\ mm$.

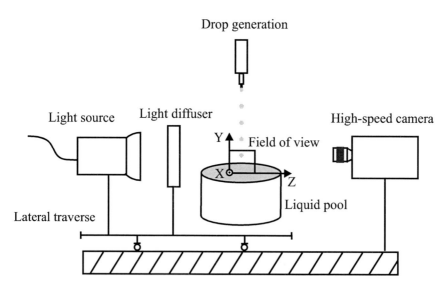

Fig. 4.1: Experimental setup for droplet impingement experimentation

In order to study the effect of impact velocity on the crown expansion, the secondary drop characteristics and the ejected mass fraction, the distance between the syringe and the target pool was changed by varying the vertical position of the syringe. The resulting range of impingement velocity was found to range from $V = 1.5\ m/s$ to $V = 3.7\ m/s$.

Target Pool

The deep liquid pool is made of fluid in a cylindrical aluminium container. The dimensions of the container had a diameter of $D = 300\ mm$ and a height of $h_f = 100\ mm$. It is large enough to avoid effects introduced by the side and/or the bottom walls. The literature review showed that during the impact of droplets onto a liquid surface capillary waves are generated at the impact point which move away from the impingement region towards the side walls of the container. In order to limit and avoid the influence of the reflected capillary waves on the impingement evolution, the diameter of the container is required to be around 100 times the largest impact drop diameter. Moreover, taking the results from Bisighini (2010) into account, a deep pool impact regime takes place when $H^* > 1.5\Delta_{max,deep}$, e.g. the liquid film height is 1.5 times larger than the maximum cavity depth. The depth of the container was chosen to be around 8 times the largest impact drop diameter for an impingement onto a deep liquid pool. Another issue that arises when using image acquisition methods for the characterization of the impingement evolution during single drop impact onto deep liquid pools is the meniscus formed by the target liquid on the walls of the container. The wall adhesion forces and the contact angle formed between the liquid and the wall lead to a meniscus that obscures the region near the free surface on the part of the recording devices and the light source, respectively. This in turn prevents the observation of the dynamics occurring in this area during the impact evolution. In order to overcome the obscured region in the field of view, the liquid level was continuously increased during the experiments until the container was completely filled. The direction of the wall adhesion forces and thereby the free surface curvature guaranteed finally a free field of view of the impact region.

Acquisition Setup

The images of the present investigation were recorded using an 8 bit CMOS sensor of a high-speed camera from Imaging Solutions GmbH, Motion Xtra N3. The camera used a Micro Nikkor Macro 105 mm f/2.8 lens. The adopted light source provided a strong back-light illumination emitted from a 1.2 KW Halogen light source and diffused through an etched glass.
The optical setup must be capable of the acquisition of the impacting drop before the impingement, the crown evolution and the secondary drops in the same field of view. The crown evolution during a single drop impingement can be estimated roughly with the help of the square-root relation of Yarin and Weiss (1995) and Cossali et al. (2004).

Consequently, the crown diameter and height are not expected to be larger than ten times the impacting drop diameter during the crown disintegration process. The lamella and crown dimensions following a drop impingement event are moreover transient. The field of view is hence selected with the dimensions of 45 mm X 35 mm. The sensor dimensions, on the other hand, are also given with a maximum size of 15.36 mm X 12.29 mm, build by 1280 px X 1024 px on a chip of 0.012 mm X 0.012 mm each. With these parameters the real dimensions of the field of view need to be magnified by

$$m = \frac{l_{sensor}}{l_{real}} = \frac{ns}{l_{real}}, \tag{4.1}$$

of 0.35, where l_{sensor} and l_{real} are the smaller size of the sensor and real field of view, respectively. s denotes the pixel width, n the number of pixel over the smaller size of the pixel length and m is called the magnification factor. Combining the pixel width and the magnification factor the spatial resolution is obtained as

$$r = \frac{s}{m}, \tag{4.2}$$

and equal to 0.035 mm/px. Assuming that 2-3 pixels are sufficient to detect the smallest drop sizes, the minimum secondary drop size resolvable is approx. $70 - 105\ \mu m$. The literature review of the previous chapters has already showed that the secondary drop sizes are approximately 10 % of the impacting drop diameter in average. By taking the adopted range of impacting drop diameters proposed in Section 4.1, the selected resolution can be seen as appropriate for the present investigation. An accurate size measurement of the impacting drop, the crown evolution and the secondary drop can moreover only be accomplished if the free surface between the liquid and the gas is sharply resolved. The free surface and the drops are then located in the plane of focus at a distance v from the lens and $u + v$ from the focal point at the sensor. By adopting the fundamental optics, the distance u of the object from the lens and the distance v of the sensor from the lens may be derived as a combination of the magnification factor

$$m = \frac{u}{v}, \tag{4.3}$$

and the thin lens equation

$$\frac{1}{f} = \frac{1}{u} + \frac{1}{v}, \tag{4.4}$$

where $f = 105\ mm$ is the focal length of the adopted lens. With this, the resulting distances are reached with $u = 141\ mm$ and $v = 405\ mm$, respectively. The former distances occur only if the objects (drops) are located in the plane of focus at one distance from the lens. All the other objects outside this plane are out of focus and appear as blur spots. Here, the greater the distance of an object from the plane of focus the larger the size of the blur spots. In practice, objects at considerably different distances from the camera can still appear sharp although not being precisely located in the plane of focus. This area/volume is often referred to as the depth-of-field (DoF). The term *sharp* is, however, of subjective value and needs therefore to be defined with a criterion. An object is of acceptable sharpness when the blur spot is indistinguishable from a point. Appreciating the fact that real lenses do not focus all rays perfectly, the term *circle of confusion* (CoC) is often used for the smallest blur spot a lens can achieve excluding the diffraction effects from wave optics and the finite aperture of a lens.

An equation defining the depth-of-field with characteristic quantities of the optical setup is proposed by Schleitzer (2002) with

$$DoF = \frac{2 \cdot NA \cdot CoC \cdot f^2 \cdot v^2}{f^4 - NA^2 \cdot CoC^2 \cdot v^2}, \tag{4.5}$$

where NA represents the numeric aperture , f, the focal length of the lens, CoC, the Circle of Confusion and v, the distance of the imaged object to the lens. The Circle of Confusion depends strongly on the optical setup, the adopted illumination, the recording device, the macro lens, diffraction effects and the finite aperture, and cannot be obtained accurately a priori. A summary of the optical settings can be found in Table 4.1.

Field of view in [mm]	45 X 35
Resolution in [$\mu m/pixel$]	35
f-number ($f/\#$)	32
Focal length in [mm]	105

Table 4.1: Characteristic parameters of the optical setup

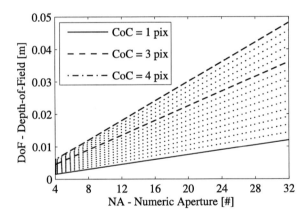

Fig. 4.2: Dependence of the depth-of-field on the numeric aperture and the circle of confusion

In Fig. 4.2, the dependence of the depth-of-field (DoF) on the numeric aperture (NA) for different circle of confusions of the present optical setup is shown. It embraces the fact that without the exact value of the CoC no exact depth-of-field size can be obtained. It is also showing that an increase in NA, hence a decrease in the light gathering area, in the optical setup leads to a larger depth-of-field. However, diffraction effects arising from wave optics set a limit for the increase of NA. These diffraction effects from wave optics exist when the light gathering area becomes to small. They can be estimated with the help of the equation defined in Hecht (2002)

$$q_1 \approx 1.22 \cdot \frac{f\lambda}{D}, \tag{4.6}$$

where q_1 is the radius of an airy disc generated by diffraction effects, D is the diameter of the aperture and λ is the wave length of the light emitted by the light source. Here, the smaller the f-number in the optical setup, the larger the diameter of the airy discs. Using this equation and the parameter of the optical setup, the diameter of the airy disc reaches approximately 60 μm. In the present investigation, the disintegrated secondary drops may be produced at different instances during the impingement evolution and may leave the impact location with different velocities and ejection angles. This implies that their location scatters in a volume around the impingement position. To ensure that most of the detected droplets are positioned within the depth-of-field, it is essential that an appropriate DoF calibration procedure will be put in place. After a brief description of the image processing technique later in this chapter in section 4.1, a DoF calibration procedure will be proposed.

The maximum frame rate of the high-speed camera with the selected dimensions of the field of view is given with $fr = 2065\ fps$. Authors like Cossali et al. (2004) and Fedorchenko and Wang (2004) showed that the crown expansion velocity, and the velocity of the secondary drops were on average in the same order of magnitude as the impingement drop velocity. Appreciating the fact that the smallest drops released by the very fast ejecta in the initial stage of the impact event are not the focus of this investigation (since they are carrying only a very small fragment of the total released mass), a sufficient capability to analyse the relevant velocities is given by the selected dimensions of the field of view.

In the case that the light source provides sufficient back illumination, the CMOS sensor resolves the light intensity with an 8 bit sensitivity or $2^8 - 1 = 255$ counts building the raw grey scale image. Intensity of light equal to or below the minimum sensitivity of the sensor returns a zero value, while when above the maximum sensitivity a value of 255 is returned. For a continuous light source, the amount of light allowed to reach the lens is controlled by the aperture of the diaphragm and by the exposure time. A change of the aperture significantly affects in turn the depth-of-field in the optical setup. Due to the large depth-of-field required in the present investigation, and setting a constant sensitivity of the sensor, the only two possible ways of increasing the amount of light reaching the lens is by selecting both a stronger light source and a larger exposure time. An increase in light intensity is limited by the heat release to the fluid and the associated change of the fluid properties during the impingement event. Larger exposure times, on the other hand, are limited by the frame rate of the high-speed camera, e.g. for a frame rate of 2065 fps the exposure time cannot be larger than 0.485 ms. The selected spatial resolution of 0.035 mm/px has also an impact on the maximum possible exposure time due to moving objects. Taking maximum drop velocities of the same order of magnitude as the impingement velocity, sharp drops can only be obtained with an exposure time not larger than 30 μs. In the present investigation a compromise was adopted with a strong halogen light source of 1.2 KW power and an exposure time of 30 μs.

Image Processing

The raw 8 bit grey scale high-speed images were processed by a *Mathworks* environment developed by Mueller et al. (2006), verified by Kapulla et al. (2008) and successfully applied by Mueller et al. (2008) and Gepperth et al. (2010). A brief description of the processing method is given subsequently.

The recorded 8 bit grey scale high-speed images were first digitalized and thus represented by a numerical matrix. Each member of the matrix corresponds to the intensity value between 0 (black) and 255 (white) of each pixel in the image. Figs 4.3a and 4.3b principally illustrate the grey-scale intensity distribution of a falling drop in a pixel row through the droplet. Here, it becomes evident that the edge of the drop is located in a region of a sharp gradient of the intensity values.

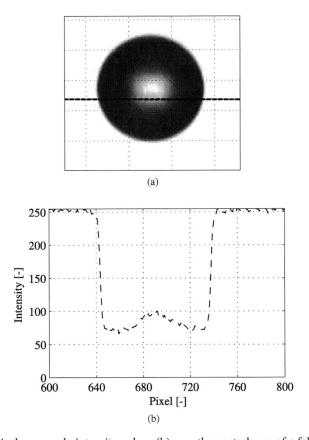

Fig. 4.3: Typical grey-scale intensity values (b) over the central row of a falling drop (a)

The Matlab image processing code is structured in four main parts. First, a background calibration image was recorded before the impingement with an equivalent setup and illumination but without any droplets in the field of view.

Then, the impacting droplet before the impingement, and the recorded secondary drops produced after the impact were separated from their background by subtracting the calibration image from the original image. Fig. 4.4 shows the background correction procedure of the original image. This step is necessary to obtain a uniform background intensity value of the images. All the intensity discontinuities of the image were then marked in a further grey-scale analysis by dividing the image into the foreground regions and the background using a threshold value with sub pixel accuracy. The detected points were subsequently interpolated and outlined with the help of polygons. Open polygons represent either particles at the border of the image or the shape of the liquid film interface structure. The closed polygons were classified as impacting drops, secondary drops and glare points in the centre of the particles. As the glare points were removed, the remaining closed polygons were fitted by spheres and ellipsoids resulting in drop sizes, positions and counts. The fitted shapes have to meet a number of conditions before being approved for further processing. These conditions are described in detail by Mueller et al. (2008). As will be shown in the next section, all the particle sizes were then evaluated and corrected based on their out-of-focus position with a depth-of-field correction procedure. In the last stage, the impacting drop and secondary drop velocity are analysed by taking two successive high-speed images. The shift of the particle centre positions is measured as a difference quotient of the particle displacement within the time interval of the two given exposures (see stage 3 in Fig. 4.5e).

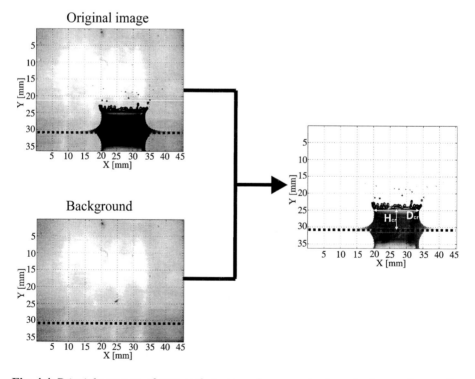

Fig. 4.4: Principle process of grey-scale image enhancement using a background correction

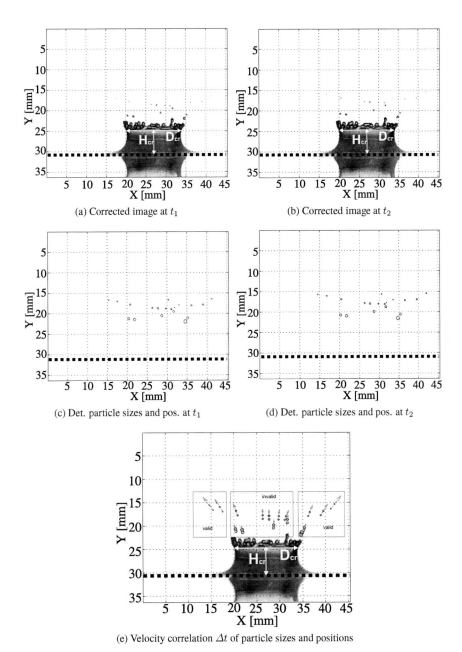

(a) Corrected image at t_1

(b) Corrected image at t_2

(c) Det. particle sizes and pos. at t_1

(d) Det. particle sizes and pos. at t_2

(e) Velocity correlation Δt of particle sizes and positions

Fig. 4.5: Principle of the post processing routine

However, the axial velocity components of the secondary drops, in particular, that are released outside of the X-Y plane are false represented in this optical setup. All the measured child drop velocities were therefore assigned to a valid zone and an invalid zone based on their positions (see Fig. 4.5e). The valid zone contains the velocity of child drops leaving the crown's rim in the X-Y plane. This velocity field can be assumed as axisymmetric during the drop impact procedure.

All other velocities were neglected in the further analysis. Fig. 4.5 summarizes the evaluation of two successive high-speed images, including background removal, detected particles (blue and red ellipses) and the calculated particle velocities (green arrows). The length of the arrow indicates the velocity magnitude while the arrow head denotes the direction of motion of the particle. The measurement uncertainties were determined by adopting the approach of Kline and McClintock (1953). The uncertainty for the velocity of the impinging drops are estimated with less than 2% whereas for the secondary drop velocities the uncertainties increased up to 4%.

Depth-of-Field Calibration

It was reported in section 4.1 that an accurate determination of the size of depth-of-field only based on the fundamental optics is not possible a priori. This is due to the dependence of the Circle of Confusion on the selected optical setup (illumination, high-speed camera, macro etc). Therefore, in this section it will be explained how to directly measure the depth-of-field of the present optical setup by adopting a calibration target. Pixel intensity gradients of different drop classes are related to their out-of-focus position in order to determine subsequently whether the measured drops are located within the depth-of-field (DoF) or not.

Contouring methods such as the one adopted in the present investigation tend to significantly over or under predict the detected drop sizes when applied to droplets outside of the DoF. In this case, the pixel intensity gradients on the edge of the droplet diminish significantly in magnitude. Hence, the drop-size measurements may be decisively false. In the present investigation, most of the particles were detected by the contouring algorithm reported in Section 4.1 by taking a pixel intensity gradient threshold of 85%. In general, such a high threshold leads to an overestimation of all the detected particle sizes. This, however, can vary depending on the drop size class. Hence, it is mandatory to examine the influence of the depth-of-field on the drop sizes and to develop suitable calibration procedures for the correction of impact and secondary drop-sizes (see Kapulla et al. (2008) and Mueller et al. (2008)). In order to examine the depth-of-field, the value of the CoC and the accuracy of the measured child drop diameter, a calibration procedure was carried out with a calibration target from $LaVision\ GmbH$. Fig. 4.6 shows this calibration target.

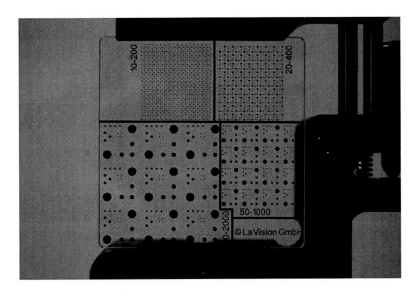

Fig. 4.6: La Vision calibration plate used for the depth-of-field calibration

A characteristic feature of the calibration target is its extensive manufacturing process. High resolution dots with very high accuracy in size and shape are located at the top of the plate at four different areas covering sizes between $10\ \mu m$ and $2000\ \mu m$, respectively. For the present investigation only one field of the calibration target was used with dot sizes between $200\ \mu m$ and $1\ mm$. Over that range, the dots should imitate the secondary drops in the later measurement. The target was mounted at a position where the later splashing drop impingement should take place. In this context, it is very important to reproduce the real impingement environment in terms of illumination and optical setup as close as possible. The 1D traverse system, as shown in Fig.4.1, has the capability to move the plate with a minimum stepwidth of $0.1\ mm$ along the z-direction thereby allowing the calibration process. A set of calibration images were recorded for every out-of-focus plane within the range of $-25mm < z < 25mm$. The positive direction refers to a movement of the calibration plate closer to the camera. Each acquired image of the calibration target in each out-of-focus planes was processed using the contouring algorithm presented in Section 4.1. Then, for every out-of-focus plane, the sizes at each dot were measured and compared to their real sizes.

Fig. 4.7 shows the resulting non-dimensional sizing accuracy for different out-of-focus positions. For the smaller dot classes, specifically for $D_{ref} < 200\ \mu m$, the over-prediction appears to be max. 20% over the range of out-of-focus planes. This over-prediction reduces continually with increasing drop size classes, as shown in Fig. 4.7.

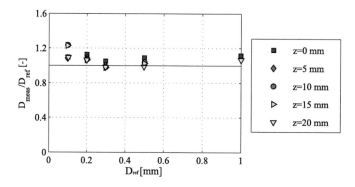

Fig. 4.7: Non-dimensional sizing accuracy for different out-of-focus positions

In order to verify later whether the detected secondary drops are located inside or outside the depth of field for each set of drop size measurement, the local intensity gradient needs also to be included in the analysis. In the post-processing procedure, the mean of the intensity gradient of every boundary pixel for each contour is calculated and normalized by the median intensity of the image. The pixel intensity gradient calculation for each dot on the calibration target in every single out-of-focus position is shown in Fig. 4.8.

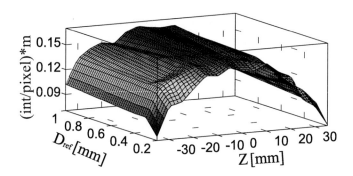

Fig. 4.8: Calibrated normalized intensity gradients as a function of z and the reference diameter

The pixel intensity gradient surface shows a significant decrease of the gradient outside the range between $z = -20\ mm$ and $z = 20\ mm$. This together with the symmetry of the gradient matrix to the out-of-focus position indicates the size of the depth-of-field. Hence, the depth-of-field of the present optical setup is obtained with approx. $DoF = 40\ mm$.

By using equation 4.5, the Circle of Confusion of the present optical setup can be calculated with approx. $40\mu m$. The latter together with the diffraction (airy disc) diameter leads to a minimum resolution of the present optical setup of around $100\ \mu m$. This means, in turn, that the minimum particle size that is detectable with the present optical setup is given with $D = 100\mu m$. An analysis of the out-of-focus position of all the measured impacting and secondary drops, based on their pixel intensity gradients, showed that all the droplets were located inside the depth-of-field. For this reason, it is not required to calibrate the diameter of the drops based on their out-of-focus position. Instead of that, three calibration functions for three different drop size classes have been derived from Fig. 4.7. The drop size classes were chosen with $D < 200\mu m$, $200 < D < 300\mu m$ and $D > 400\mu m$. The functions were determined for an average of all the over-predicted sizes along the z-direction. Using these calibration functions, the measurement uncertainties for the drop sizes resulted in a 4% uncertainty in drop size for $170\mu m < D < 200\mu m$, in 3% for $200\mu m < D < 300\mu m$ and in below 2% for all the larger sizes.

4.2   Results and Discussion

The experimental validation of the 2D axisymmetric and the 3D VOF-AMR method is divided into two parts. The first part compares the results of the 3D-VOF-AMR for the crown's disintegration during single drop impact with the measurements conducted by the experimental method described in the beginning of this chapter. Here, the quantities that will be compared to the numerical predictions are the characteristics of the secondary droplets and the ejected mass fraction. In the second part of the experimental validation, the 2D-VOF-AMR method is validated using high-quality experimental data from the scientific literature. For the validation of both approaches, large millimetre sized drop impingements onto a deep liquid pool are selected since only these time and length scales can be accurately accessed with the available experimental methods. For all the investigated impingement conditions, the test fluids of the impinging droplet and of the target pool were the same for every impingement event.

The liquid film height, on the other hand, was selected on a case by case basis to avoid additional wall effects. The impinging Weber number was varied with the impingement velocity and/or the diameter. Impingements of higher viscous fluids were achieved with the aid of glycerine.

4.2.1   Study of the Transitional Threshold between Spread and Splash

Before assessing the numerical methods, a brief description of the identified transitional threshold between spread and splash is given in this section. Here, the objective is to cover the very large range of liquid viscosity occurring in bearing chambers and compare the thresholds with available correlations in scientific literature. The prompt and corona splash boundary is determined using single drop impacts onto a deep liquid pool. The test cases are summarized in table 4.2. Liquid viscosity is varied using different water/glycerin mixtures that corresponds to a range of Ohnesorge number between $0.0018 < Oh < 0.036$. On the other hand, the range of impinging Weber numbers results between $50 < We < 1300$. Fig. 4.9 shows a We-Oh map containing all the investigated experiments in the present thesis. The quads refers to those impingements resulting with the splashing regime whereas the dots represent the spreading regime. With respect to the splash regime both prompt and corona splashing are described. The comparison of the

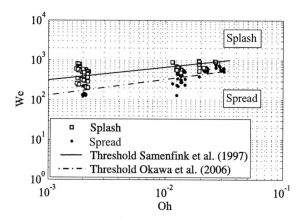

Fig. 4.9: Various splashing limits for single drop impact onto shallow and deep liquid pool
($H^* > 1$)

transitional thresholds with the threshold $K_{crit} = 2100$ of Okawa et al. (2006) and Rioboo et al. (2002) shows a very good agreement for low Ohnesorge numbers, between $0.001 < Oh < 0.003$. However, only a fair comparison exists for the larger viscosity impingements since their threshold is under-predicting the initiation of splashing identified in this thesis. In contrast, the correlation of Samenfink (1997) predicts the splashing regime at much larger Weber numbers than found with the present results. Although it is not indicated in their publication, it appears that only the corona splashing was defined as the splashing regime.

In conclusion of this subsection, it may hence be stated that the transitional threshold determined by Okawa et al. (2006) and Rioboo et al. (2002) appears to be valid also for the description of splashing formation in this thesis. Other correlations from scientific literature appeared to significantly under or over-predict the initiation of splashing in particular for impingements with very large Ohnesorge numbers (see Vander Wal et al. (2006a)).

4.2.2   Assessment of the 3D-VOF-AMR Method

A very challenging task, to be discussed now, is the validation of the 3D-VOF-AMR for the direct simulation of the crown's breakup during single drop impingement. The secondary droplet generation due to the corona breakup during single drop impingement is known to be a fully three-dimensional phenomenon despite the mostly axisymmetric nature of the cavity and lamella rim evolution to be presented in the next section. It therefore requires a 3D solution of the Navier-Stokes equations coupled with a VOF method and an adaptive mesh refinement technique. The experimental validation of the 3D-VOF-AMR method was conducted qualitatively and quantitatively using the high-speed shadowgraphs and the measurements of sizes, velocities, and masses of the secondary drops produced by single drop impacts onto deep liquid pools. However, before leading to the experimental validation, for completeness, a brief discussion of the qualitative and quantitative experimental results of secondary drop generation during a deep pool impingement will be presented.

The impingement conditions for analysing the secondary drop characteristics and ejected mass fraction with the 3D-VOF-AMR method are listed in table 4.2. XX in the terms $WGXX$ in Table 4.2 refers to the total mass fraction of glycerin added to water to increase the liquid viscosity. Each impingement event is repeated five to ten times improving thereby the statistics. The impingement conditions were selected to ensure the splashing regime.

case	Fluid	D [mm]	V [m/s]	We	Oh	Fr
1	Water	2.9	2.94	353	0.0022	303
2	Water	2.9	3.61	529	0.0022	458
3	Water	4.2	3.7	800	0.0018	332
4	Water	2.8	4.2	683	0.0018	642
6	WG45	4.0	3.7	880	0.011	348
7	WG65	3.9	3.7	880	0.027	357

Table 4.2: Test cases for validation of the 3D-VOF-AMR method

Qualitative Comparison of the Impingement Outcome

Figs 4.10a-4.10f show the temporal evolution of a water drop impingement onto a deep liquid pool at six different time instances. Here, the dashed line represents the free surface of the liquid pool. The impingement condition uses an impacting droplet diameter of $D = 4.2\ mm$ and an impacting velocity of $V = 3.7\ m/s$, respectively. In the initial stage of the impact scenario, for $\tau < 0.25$, a prompt splash regime takes place.

In the further evolution of the impact, $0.25 < \tau < 8$, a much slower and thicker lamella emerges and increases in diameter and height with time until it reaches a maximum height. During the crown expansion, the corrugations at the rim of the lamella grow in magnitude until a point where finger-like jets again protrude from the rim and break up into secondary drops, as seen in Fig. 2.5c. With increasing non-dimensional time, the thickness of the finger-like jets increases while the length reduces. Hence, the size of the secondary drops released due to the jet-breakup of the finger-like structures is also increasing the impact time. This was also observed and measured by Cossali et al. (2004). After reaching its maximum height and diameter, the lamella stops its expansion and collapses, between $8 < \tau < 12$. Gravitational forces of the lamella dominate the inertial forces, and the corrugations at the rim of the crown disappear rapidly. For $\tau > 25$ an axisymmetric Rayleigh-jet evolves perpendicularly to the free surface, leading to a droplet pinch-off. According to Macklin and Metaxas (1976) and Okawa et al. (2006), the generation of secondary drops from the crown's rim for single drop impacts onto deep liquid pools was

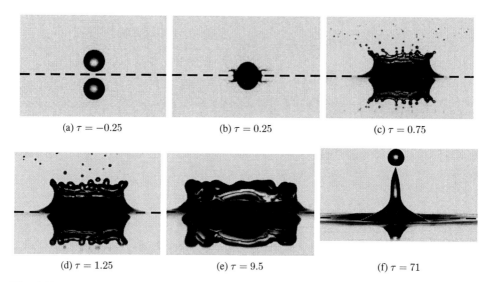

(a) $\tau = -0.25$ (b) $\tau = 0.25$ (c) $\tau = 0.75$

(d) $\tau = 1.25$ (e) $\tau = 9.5$ (f) $\tau = 71$

Fig. 4.10: Drop impact onto a deep liquid pool: Role of time during the impact evolution (We=800, Oh=0.0018)

observed only in the expansion phase of the crown. While collapsing and falling back, none or only occasionally some larger secondary drops were generated. This is in contrast to the impact of droplets onto thin liquid layers, where large drops are also generated during the corona collapsing stage (see Cossali et al. (1997), Rioboo et al. (2003), Cossali et al. (2004), Vander Wal et al. (2006a), and van Hinsberg (2010)).

A qualitative comparison of the crown's formation, the corona breakup and collapse regime is depicted in Fig. 4.11. Here, the left hand side of the figures represent the 3D-VOF-AMR results by the iso-surface of the volume fraction at $\alpha = 0.5$ while the right hand side depicts the high-speed shadowgraphs of the experimental setup described in Section 4.1. The corresponding non-dimensional times are also indicated. The impingement condition corresponds to case 3 in Table 4.2. In the figures, the black line represents the undisturbed liquid film surface. In order to provide an improved visualization of the crown's breakup dynamics, all figures were magnified appropriately.

The comparison of the numerical results with the experimental high-speed recordings proves the good representation of the crown's dynamics at all occurring breakup stages. Both the complex ejecta breakup in the prompt splashing regime (at $\tau = 0.25$) and the sheet disintegration in the corona splashing regime (between $0.25 < \tau < 2.25$) are in excellent agreement with the (high-speed) recordings.

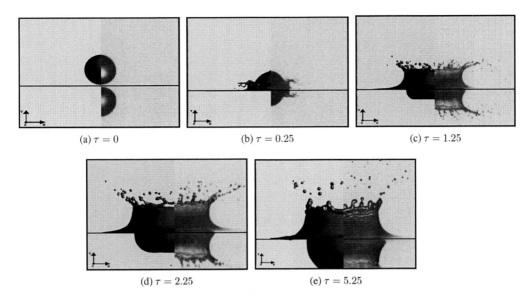

(a) $\tau = 0$ (b) $\tau = 0.25$ (c) $\tau = 1.25$

(d) $\tau = 2.25$ (e) $\tau = 5.25$

Fig. 4.11: Comparison of a 3D-VOF-AMR simulation (left: iso-surface at $\alpha = 0.5$) with shad-owgraphs (right) at different impingement stages: water, $D = 4.2\ mm$, $V = 3.7\ m/s$ (case 3)

The number of corrugations at the rim, the protrusion of finger-like jets, and the breakup into secondary drops are correctly captured by the numerical method. However, the number of secondary drops seems to be somewhat smaller than in the experiment. This is caused by an insufficient spatial resolution, which would cause the smallest secondary droplets, to be not captured. As a consequence, the secondary drop sizes are slightly over-predicted. Their calculated ejection angle, on the other hand, agrees well with the shadowgraphs (see Fig. 4.11b-4.11e). On some occasions, when the rim constricts due to surface tension, within the calculation holes are simulated underneath the rim which are not observed in the experiments. This shortcoming of the numerical method was also observed by Rieber and Frohn (1999) in their simulations and can also be explained by an insufficient local mesh resolution. However, no major influences on the secondary drop characteristics and ejected mass fraction were identified due to the former inconsistency.

The best agreement between the simulation and the shadowgraphs is given for the corona collapse phase at $\tau = 5.25$ (see Fig. 4.11e). At this stage, nearly every detail is captured despite the slight over-prediction of the secondary drop sizes. Another feature that appears less pronounced in the simulation is the propagation of large capillary waves along the lamella rim, continuously perturbing the crown. The 3D-VOF-AMR method simulates the capillary waves with a corresponding wavelength but considerable smaller amplitude than in the experiments. Nevertheless, the numerical method is capable of qualitatively resolving the crown's breakup process. This fact needs to be particularly emphasized since most of the numerical methods in the scientific literature were not able to resolve the protrusion of finger-like jets from the crown's rim.

Comparison of the Secondary Droplet Characteristics

The data indicated in Fig. 4.12 illustrate a typical experimental result showing the characteristics of the secondary droplet motion following the normal impact of a water drop onto a deep liquid pool for impact conditions of $D = 4.2\ mm$ and $V = 3.7\ m/s$. These impact conditions ensure an extensive disintegration of the crown's rim and are selected for further analysis.

Fig. 4.12a shows the relative and cumulative number distributions of all the secondary drop sizes detected over the whole disintegration process excluding the large drop pinch-off. In the very early stages of the normal splashing impact, an ejecta forms and releases several micrometric secondary drops that were not resolved by the optical setup adopted (due to the selected pixel resolution). This fact is also evidenced in the relative and cumulative number distributions since no secondary droplets were detected with sizes smaller than $D_{sec} < 170\mu m$. A peak in the relative number distribution is noted at a secondary drop size of $D_{sec} = 310\mu m$ corresponding to the arithmetic mean diameter. Fig. 4.13 depicts the ejected mass carried by every single secondary droplet generated during a typical single drop impingement onto a deep liquid pool. Here, it is evident that nearly 70% of the mass is carried by the 5 largest secondary drops, out of 74.

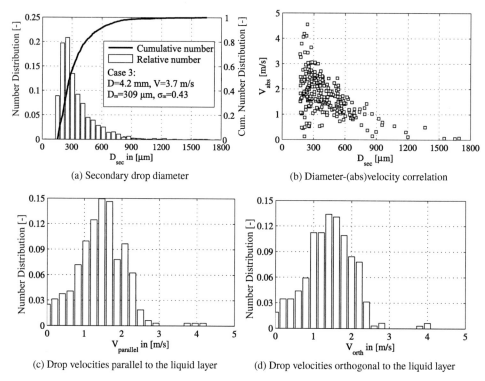

(a) Secondary drop diameter

(b) Diameter-(abs)velocity correlation

(c) Drop velocities parallel to the liquid layer

(d) Drop velocities orthogonal to the liquid layer

Fig. 4.12: Motion of secondary drops in the impingement region during a splashing water drop impingement onto a deep liquid pool: $D = 4.2\ mm$ and $V = 3.7\ m/s$

The existence of relatively large secondary drops up to $D_{sec} \approx 1600 \mu m$, carrying the major part of the released mass, is also indicated in the relative and cumulative number distributions in Fig. 4.12a. The major advantage of the high-speed image processing technique is the accurate measurement of the large secondary drops, which are mostly non-spherical and therefore difficult to detect with phase Doppler anemometry. These large drops are produced in the final disintegration stage of the corona when the gravitational forces drive the motion of the thick corona back towards the liquid film.

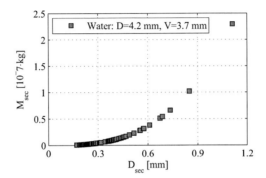

Fig. 4.13: Typical relation between the secondary drop mass and secondary drop diameter for a water drop impingement of $D = 4.2\ mm$ and $V = 3.7\ m/s$

In Fig. 4.12b, the size-(abs)velocity correlation for each detected secondary drop during the splashing impingement is depicted. The small drops released from the rapidly expanding crown, in the initial stage of the impingement process, are released with velocities that can be larger than the impingement drop velocity. The gravitational and surface tension forces acting at the crown's rim at this time is not large enough to moderate the absolute velocity of the secondary drops. With increasing impingement time, more and more mass flows into the crown's rim, leading to an increase in thickness (potential energy) and capillary forces (see, for example, Cossali et al. (2004)). Hence, large secondary drop sizes are disintegrating at very low velocities, as is visible in the size-velocity correlation. Stow and Stainer (1977), Samenfink (1997) and Cossali et al. (2004) confirmed this scenario for single drop impacts onto wetted surfaces, and onto both thin and shallow liquid films. In order to gain more knowledge about the ejection angles of the secondary drops, the absolute velocity is decomposed into components parallel and orthogonal to the undisturbed liquid layer surface and plotted with the relative number distribution in Figs 4.12c and 4.12d. The peaks of the relative number distribution occurs at $V_{parallel} = V_{orth} \approx 1.8$ m/s for both secondary drop velocity components, resulting in a peak ejection angle of around $\alpha_{sec} \approx 45°$. The smaller drop sizes leave the impingement point with decreased orthogonal and larger parallel secondary drop velocities, hence with flatter ejection angles, and vice versa for the larger drop sizes.

At this stage, it is convenient to reduce the number of parameters needed to entirely describe the secondary drop size distribution obtained. This is usually achieved by adopting mathematical distribution functions. These have the advantage of describing an entire (size) distribution

with only a limited number of parameters. An extensive review on this topic is provided by
Babinsky and Sojka (2002). In chapter 2, most of the secondary drops produced from single
drop impingement onto wetted surfaces, thin and shallow liquid film could be successfully fitted
by log-normal distribution functions (see Levin and Hobbs (1971), Stow and Stainer (1977),
Stow and Hadfield (1981), and Samenfink et al. (1999)). The log-normal probability distribution
function defined as

$$f_0(D) = \frac{1}{D(ln\sigma_m)\sqrt{2\pi}}exp\left\{-\frac{1}{2}\left[\frac{ln(D/D_m)}{ln\sigma_m}\right]^2\right\},$$

(4.7)

where D is the secondary drop diameter, σ_m the width of the distribution, and D_m the arithmetic
mean diameter. A cumulative distribution function is then simply obtained by integrating Equation
4.7. The arithmetic mean diameter D_m is commonly obtained by using the expression developed
by Mugele and Evans (1951) with

$$D_{pq} = \left[\frac{\sum_{i=1}^{N} D_i^p}{\sum_{i=1}^{N} D_i^q}\right]^{\frac{1}{(p-q)}},$$

(4.8)

where N denotes the number of detected child drops. The arithmetic mean diameter D_{10} is
obtained with $p = 1$ and $q = 0$. Fig. 4.14 compares the secondary drop size distribution obtained
from a water drop impact onto a deep liquid pool with the log-normal probability and cumulative
density function for the respective arithmetic mean diameter $D_m = 309\mu m$ and the distribution
width of $\sigma_m = 0.43$. Both the probability and the cumulative density functions in Figs 4.14a and
4.14b reveal a good match for the number, area, and volume distribution of the secondary drop
sizes, and thus support the applicability of the log-normal distribution function.

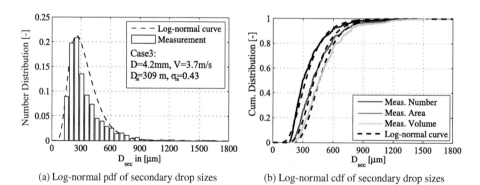

(a) Log-normal pdf of secondary drop sizes   (b) Log-normal cdf of secondary drop sizes

Fig. 4.14: Fitting of a log-normal distribution function onto the measured secondary drop diameter distribution of a splashing water drop impingement onto a deep liquid pool: $D = 4.2$
mm and $V = 3.7 \, m/s$

In order to allow a detailed comparison of the measured secondary drop data with predictions existing in the scientific literature, the arithmetic mean secondary drop size and velocity are plotted versus the K number in Fig. 4.15. Here, in the Figs 4.15a and 4.15b, the arithmetic mean secondary drop size and distribution width of all the investigated test cases are depicted and compared to the empirical correlations obtained by Roisman et al. (2006) for sparse spray impact onto thin liquid films, Samenfink et al. (1999) onto shallow liquid films, and Okawa et al. (2006) onto deep liquid pools. In general, an increasing K number causes a weak increase in the mean secondary drop size in the present investigation that is not confirmed by the trend identified by Roisman et al. (2006) and Samenfink (1997) but appears to be in very good agreement with the findings of Okawa et al. (2006).

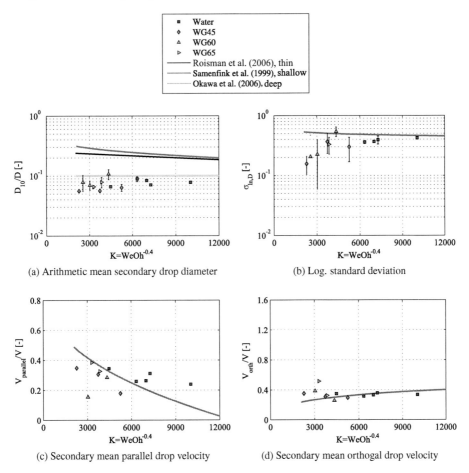

Fig. 4.15: Measured arithmetic mean diameter and velocity of secondary drops versus the K number of all the investigated test cases and comparison with available correlations in the scientific literature

Note that the predictions of Samenfink et al. (1999) and Roisman et al. (2006) delivered only reasonable results for low Ohnesorge numbers. For higher viscous fluids the predicted secondary droplet sizes were significantly overestimated. This may be related to the fact that only water was used as test fluid (narrow validity). Moreover, the existence of a deep liquid pool during drop impingement compared to the thin and shallow impacts, and the strong difference in impinging drop sizes (mm vs μm) already indicate that there is a change in the impingement dynamics. These differences will be described in detail later in this thesis using the Direct Numerical Simulations.

The arithmetic mean absolute velocities are divided into two components, one parallel and one orthogonal to the undisturbed liquid surface, and plotted versus the K number in Figs 4.15c and 4.15d. The parallel and orthogonal secondary drop velocity appears to be in good agreement with the correlations developed by Samenfink et al. (1999) for shallow shear driven liquid film impingements.

A comparison of the measured and calculated 3D-VOF-AMR local secondary drop sizes produced by a single drop impingement onto a deep liquid pool is depicted in Fig. 4.16. The impingement condition corresponds again to case 3 (see Table 4.2). The comparison shows not only a rather good agreement of the width of the size distribution but also of the arithmetic mean diameter. A slight over-prediction of the measured secondary drop sizes occurs throughout the different size classes.

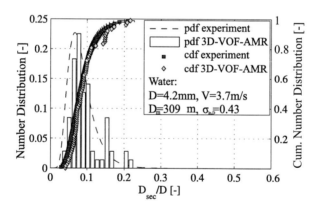

Fig. 4.16: Comparison of the calculated number distribution of the secondary drop sizes during single drop impingement onto a deep pool with the experimental results

A more general comparison between the numerical and experimental arithmetic mean secondary drop diameter and velocities is presented for different impingement conditions in Figs 4.17a and 4.17b. The arithmetic mean values are plotted here versus the impinging K-number, $K = WeOh^{-0.4}$. The mean secondary drop sizes predicted by the correlation of Okawa et al. (2006) are further delineated in Fig. 4.17. Here, the diamonds representing the simulations correspond well to the experimental measurements of the arithmetic mean secondary drop sizes.

In general, the numerical simulations show less scatter in the data than the experimental results. The numerically obtained secondary drop mean diameters also compare well with the correlation of Okawa et al. (2006) showing a constant $D_{sec}/D \cong 0.1$ over the whole range of K although showing lower values of D_{sec}/D for $K < 7000$.

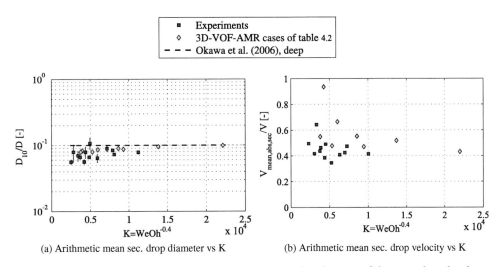

(a) Arithmetic mean sec. drop diameter vs K   (b) Arithmetic mean sec. drop velocity vs K

Fig. 4.17: Comparison of the calculated size and velocity distributions of the secondary droplets during single drop impingement onto a deep pool with experimental results and the correlation of Okawa et al. (2006)

The predicted secondary drop velocities, in contrast, are larger than the experimental measurements over the entire range of K numbers. However, a clear falling trend of the calculated values with increasing K is seen in 4.17b. Many researchers, e.g. Mundo et al. (1995) and Samenfink et al. (1999), present also in their measurements a decrease of the absolute secondary drop velocity with increasing drop impact momentum. The limitations arising from the present optical setup with respect to the secondary drop velocity measurement could explain the discrepancy between experiment and simulation.

Comparison of the Ejected Mass Fraction

The mass fraction ejected following drop impingement is considered to be the most important target parameter. It is the ratio between the ejected mass of the secondary drops and the impinging drop mass. Other definitions, such as, e.g. the deposition rate in Samenfink et al. (1999), can be found in the existing literature. This definition can be principally derived from the ejected mass fraction. Typical experimental measurements of the ejected mass fraction are presented along with the measured number of secondary drops N in Fig. 4.18. A comparison of the measured number of secondary drops and the ejected mass fraction during a single drop impact onto a deep liquid pool with the correlations available in the literature is again presented by plotting both quantities versus the K number in Figs 4.18a and 4.18b. A nearly linear increase of the ejected mass fraction and the number of secondary drops with the K number is visible. The former supports once more the applicability of the quantity $WeOh^{-0.4}$ for the description of drop impingements onto a deep liquid pool with low Froude numbers. The proposed correlation for the ejected mass fraction of Okawa et al. (2006) with $M_{sec}/M = 0.00156e^{0.000486K}$ gives a rather good agreement with the present results up to $K \approx 7000$. From here on, a significant over-prediction of the ejected mass fraction is found. This discrepancy may be explained by the missing role of the liquid film height identified in their investigation. The correlation for the ejected mass fraction appears to be completely independent of the liquid film height. This appears questionable since the validity of the correlations was claimed for impingements ranging from very thin films to deep pool impacts $0.14 < H^* < 64$. The thin film impingements, in particular, are known to be significantly affected by the wall, as already demonstrated in Chapter 2 for $H^* < 1$. Hence, the experimental results of Okawa et al. (2006) for the ejected mass fraction for $K > 7000$ were derived from experiments that contained a considerable influence of the wall which is not reflected by the correlations. Moreover, the considerably larger ejected mass for impacting K number $K > 7000$ can be justified since the main data of Okawa et al. (2006) for $K > 7000$ was combined with single drop impact events onto thin liquid layers (e.g. with the experiments of Stow and Stainer (1977)).

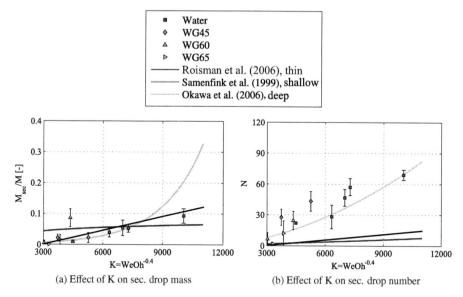

(a) Effect of K on sec. drop mass   (b) Effect of K on sec. drop number

Fig. 4.18: Effect of the We, the Oh and the K number on secondary drop mass and number during single drop impingement on deep liquid pools and comparison with available correlations in the literature

The correlation of Roisman et al. (2006) and Samenfink et al. (1999) appear to be in the same order of magnitude as the present results, with a slight over-prediction of the ejected mass for $K > 7000$.

The experimental results for the number of secondary drops agree well with the correlation of
Okawa et al. (2006) (see Fig. 4.18b). The correlations of Roisman et al. (2006) and Samenfink
et al. (1999), on the other hand, considerably under-predict the experimentally measured number
of secondary drops. The predicted number of secondary drops of both is very low for the investigated impingement conditions, as was also noticed in the review of Cossali et al. (2005). This fact
may be related to an inconsistency of the applied phase Doppler measurement technique for large
(non-spherical) drop sizes. In their investigation, the number of secondary drops was evaluated
using the volumetric mean diameter D_{30} and the ejected mass. Moreover, the predictions of
Roisman et al. (2006) and Samenfink et al. (1999) again showed for both the ejected mass fraction
and the number of secondary drops to deliver reasonable results only for low Ohnesorge number.

In the Figs 4.19a and 4.19b, the simulated ejected mass fraction M_{sec}/M and the number of
secondary drops N are plotted versus K and compared to the experimental test cases summarized
in Table 4.2. The correlation of Okawa et al. (2006) is also delineated. Here, a very good
agreement between the numerical and experimental results is given (see Fig. 4.19). However, as
a consequence of the small over-prediction of the secondary drop sizes, a weak over-prediction
occurs also for the ejected mass of secondary drops.

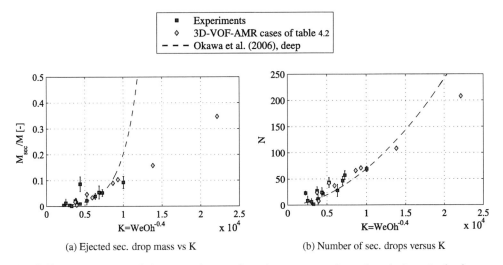

(a) Ejected sec. drop mass vs K   (b) Number of sec. drops versus K

Fig. 4.19: Comparison of the ejected secondary drop mass and number during single drop
impingement onto a deep pool with the experimental results of chapter 4 and the
correlation of Okawa et al. (2006)

4.2.3   Assessment of the Axisymmetric 2D-VOF-AMR Method

The experimental and analytical validation of the axisymmetric 2D-VOF-AMR simulation technique of the cavity-wall interaction and lamella rim evolution will be discussed using the experimental results of Liow (2001) and Bisighini (2010), combined with the experimental results produced with the experimental setup described at the beginning of this chapter. The test conditions of the experimental validation are summarized in Table 4.3.

case	Fluid	$D\,[mm]$	$V\,[m/s]$	We	Oh	Fr	Reference
1	Water	2.1	1.9	105	0.0026	173	Cavity: Liow (2001)
2	Water	2.1	2.8	236	0.0026	380	Cavity: Liow (2001)
3	Water	2.1	3.5	371	0.0025	594	Cavity: Liow (2001)
4	Water	2.2	2.4	170	0.0020	266	Cavity: Bisighini (2010)
5	Water	2.2	3.6	406	0.0020	600	Cavity: Bisighini (2010)
6	Water	2.8	4.2	690	0.0018	642	Cavity: Bisighini (2010)
7	Water	4.3	2.4	360	0.0018	136	Crown: Own
8	Water	4.3	3.1	580	0.0018	227	Crown: Own
9	Water	4.2	3.7	800	0.0018	332	Crown: Own

Table 4.3: Test cases for validation of the axisymmetric 2D-VOF-AMR method

A qualitative comparison of the high-quality shadowgraphs of Bisighini (2010) at different time stages with the numerical simulation results of the present thesis is shown in Fig. 4.20. The simulation result is depicted using the iso-line of the volume fraction at $\alpha = 0.5$. The temporal sequences clearly evidence the good agreement of the crater and lamella shape during the formation and expansion phase. The crater takes a nearly hemispherical shape during the expansion stage. A good agreement exists also in the cavity retraction stage mainly driven by capillary forces and waves that are clearly visible in Figs 4.20d and 4.20e. The calculated capillary waves are slightly more pronounced in the numerical simulations than in the shadowgraphs. Moreover, the Worthington jet building up in the end phase of the impingement tends to initiate sooner than in the experiments. This behaviour was also observed by van Hinsberg (2010). He related this behaviour to three-dimensional effects that are not resolved by the axisymmetric numerical predictions.

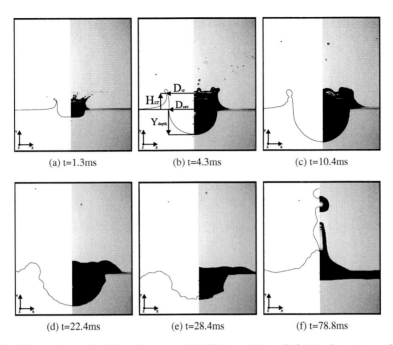

(a) t=1.3ms   (b) t=4.3ms   (c) t=10.4ms

(d) t=22.4ms   (e) t=28.4ms   (f) t=78.8ms

Fig. 4.20: Comparison of a 2D axisymmetric VOF simulation (left: iso-line at $\alpha = 0.5$) with shadowgraphs (right) of Bisighini (2010) at different impingement stages (see case 6 in Table 4.3 for details)

Fig. 4.20f underlines the former statement, showing the more elongated jet and the sooner drop pinch-off of the numerical simulations. Although not capturing the three-dimensional rim perturbations leading in some occasions to the crown's breakup, a surprisingly good agreement can be confirmed for the lamella rim evolution at different time stages. Hence, three-dimensional effects on the cavity and lamella rim evolution are expected to be small, and the methodology presented seems suitable for capture this mechanism. Thus, it can be used for qualitatively analysing the cavity and crown evolution for the range of impingement conditions found in the bearing chamber.

A quantitative validation of the two-dimensional axisymmetric VOF-AMR method is attempted next by comparing both the temporal evolution of the cavity depth Y_{depth} and diameter D_{cav}, the lamella rim height H_{cr} and diameter D_{cr} with the corresponding experimental data and results of analytical models. In order to gain scalable results, all parameters are described in a non-dimensional form. For instance, the drop impingement diameter D is used to scale the non-dimensional cavity depth $\Delta = Y_{depth}/D$, the shape factor $e = Y_{depth}/D_{cav}$, the non-dimensional lamella rim height H_{cr}/D, the non-dimensional lamella rim diameter D_{cr}/D, and the non-dimensional time $\tau = tV/D$.

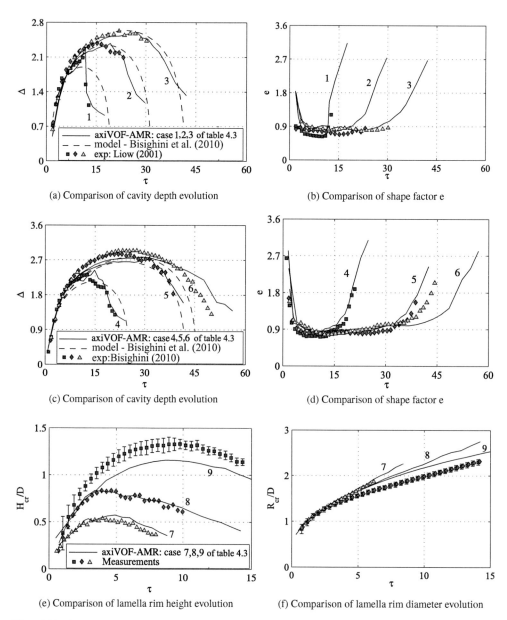

(a) Comparison of cavity depth evolution

(b) Comparison of shape factor e

(c) Comparison of cavity depth evolution

(d) Comparison of shape factor e

(e) Comparison of lamella rim height evolution

(f) Comparison of lamella rim diameter evolution

Fig. 4.21: Comparison of cavity and lamella rim evolution with experimental results of Liow (2001), Bisighini (2010) and Chapter 4 (see details in Table 4.3)

For a correct comparison with experimental results, it is moreover essential to measure the respective values of the parameter of interest at corresponding locations. In the shadowgraphs, the cavity depth is measured at the lowest point of the cavity from the selected perspective, whereas in the simulations, it is extracted at the lowest computational cells where the volume fraction is $\alpha = 0.5$. The cavity diameter, on the other hand, is measured in the experiments as close as possible to the undisturbed liquid surface (capillary rise of the liquid layer meniscus) and directly at the undisturbed free surface in the numerical simulations. This discrepancy adds additional uncertainties into the measurement of the cavity diameter. The lamella rim height and diameter are extracted at the base of the rim, whenever a rim with liquid accumulation due to surface tension forces could be clearly distinguished from the rising cylindrical lamella rim. In Fig. 4.20b, the exact measurement locations of the different parameters are depicted.

Plots of the cavity depth and the shape factor versus the non-dimensional time are illustrated in Fig. 4.21 for different impingement Weber numbers and compared to the experimental results of Liow (2001) and the analytical model of Bisighini et al. (2010), respectively (see case 1-6 in Table 4.3). In general, the numerical simulations, experimental results, and analytical model agree well. The analytical model appears to be accurate for larger We and Fr numbers where the capillary forces play a minor role. In cases where the capillary forces are dominant, as e.g. in cases 1 and 4 of Table 4.3 plotted in Figs. 4.21a and 4.21c, the analytical model deviates in resolving particularly the retraction phase of the cavity. There are some circumstances, however, and particularly in cases 5 and 6 of Table 4.3 plotted in Fig. 4.21c, where an under-prediction of the experimental results occurs on the part of both the numerical and analytical models. Nevertheless, the trends of the experimental results are very well represented by the present numerical method. Another interesting fact, apparently not reflected by the experimental curves, is the bubble entrainment taking place for some test cases (cases 1,2,4, and 5 of Table 4.3) at maximum cavity depth expansion, showing a step-wise increase and decrease of the cavity depth in time. This phenomenon, however, takes place within very short time intervals, often not resolved by the high-speed imaging techniques adopted. A comparison of the non-dimensional lamella rim height and diameter evolution in time with the new experimental results presented in Chapter 4 is furthermore depicted in Fig. 4.21 for three different impingement Weber numbers (see cases 7,8, and 9 in Table 4.3). Here, the numerical method appears to capture the lamella rim height expansion very well for the two lower Weber numbers, but under-predicts the highest Weber number by approximately 20%. The reason for this under-prediction may well be assigned to the axisymmetric nature of the numerical method, which does not resolve the three-dimensional crown breakup dynamics. Considering the fact, however, that only an axisymmetric VOF-AMR method is used for the determination of the lamella rim expansion, the results can be considered as sufficiently accurate compared to other numerical models. The evolution of the lamella rim diameter in Fig. 4.21f, on the other hand, seems to be very well resolved by the numerical method with some slight over-predictions towards the later expansion stages.

4.3   Concluding Discussion

In conclusion, an improved insight has been gained into the products of splashing using the enhanced high-speed shadowgraphy and PTV technique proposed in Section 4.1. The experimental results obtained in the present investigation have shown that the existence of a deep liquid pool changes considerably the dynamics of the products of splashing, compared to other impingement regimes, such as wetted walls-, and thin and shallow impacts. The proposed 3D-VOF-AMR method is suitable for the direct simulation of the complex three-dimensional disintegration process of the crown during single drop impingement onto liquid surfaces. Experimentally measured local secondary drop sizes appear to be well reproduced by the numerical method, together with the global ejected mass and the number of the secondary drops, over a wide range of We and Oh numbers. The most recent correlations of Okawa et al. (2006) predict both the mean secondary drop sizes of the numerical simulations and experimental results with acceptable accuracy. The correlated ejected mass and number of secondary drops of Okawa et al. (2006), in contrast, delivers very large (non-physical) values for $K > 7000$, which cannot be confirmed by the present numerical and experimental results. The absolute arithmetic mean secondary drop velocity is over-predicted by the numerical method and, moreover, decreases with increasing K numbers, in contrast to the experimental values. The over-prediction is however still acceptable since it may be related to experimental uncertainty. With respect to the two-dimensional axisymmetric VOF-AMR technique, it can be concluded that the method predicts the crater-wall interaction and the lamella rim expansion during the impingement of single droplets onto liquid surfaces. Hence, both the 2D and 3D method have shown to work accurate, and robust and with considerably lower computational effort than previous studies.

Despite the valuable insights obtained with the new experimental and numerical results gained up to this point, there are still some effects that need to be understood for a complete formulation of a drop-film interaction model for bearing chamber impingement conditions. For instance:

- How do the impingement dynamics of millimetre sized drop impacts change with drop impacts in the sub-millimetre range for an equivalent K number?

- How does a change of impingement angle affect the products of splashing during submillimeter drop impingement onto deep liquid pools?

- What additional influences occur for impact events onto shallow liquid layers?

The limitations arising from the experimental measurement techniques have been shown to make it very difficult to find appropriate answers to these questions. The numerical method will therefore be used in the next chapters to study the influence of several parameter on the single drop impingement dynamics with and without wall-effects for conditions differing significantly from the millimetre sized drop impingements usually found in the scientific literature. With the numerical database produced, correlations derived from the direct numerical simulations will be developed for the parameter range found in bearing chambers.

5 Products of Splashing Drop Impingements without Wall Effects

After the successful experimental validation of the 2D-VOF-AMR and 3D-VOF-AMR methods in the last chapter, an application of the validated numerical methodology to the drop impingement conditions relevant to aero-engine bearing chambers is presented in this chapter.

In the beginning of this chapter, a general description of a droplet impingement in the splashing regime is given. This is done qualitatively by characterizing the simulated temporal evolution of the crown's breakup and quantitatively by studying the wavelength selection of the instability driving the crown's sheet breakup. The latter is compared to traditional and recent theoretical studies. The numerical investigation continues with a parameter study where the influence of the We, Fr, Oh, α on the products of splashing is systematically studied by (initially) neglecting the wall-effects.

5.1 Fundamental Description of a Splashing Impingement

The general description of the numerically calculated mechanism leading to the breakup of the crown's rim is accomplished with the help of an impingement of a single drop on a liquid film without the influence of the wall. The impingement conditions were selected to be with $We = 2209$, $Oh = 0.0031$ and $Fr = 343$. The impinging K-number for the underlying impingement is $K = 22266$, hence, far above the critical K-number of $K_{crit} = 2100$ for the production of secondary drops due to prompt splashing and delayed corona breakup (see Section 2.5).

A temporal evolution of the crown's breakup is depicted in the Figs 5.1 and 5.2 using the iso-surface of the volume fraction at $\alpha = 0.5$ viewed from two different perspectives. Fig. 5.1a shows the single droplet and the undisturbed liquid surface before the impingement. When the single drop touches the liquid surface with a point contact, capillary waves are induced by the numerical method. These capillary waves travel upwards along the impinging drop and flow along the fast expanding ejecta (not shown in Figs 5.1 and 5.2). The former is also confirmed by the findings of Wang et al. (2002). The ejecta grows to a lamella and expands in diameter and height. The surface tension forces, which are dominant at the rim of the lamella, cause a local accumulation of liquid at the rim. This torus-like liquid accumulation becomes unstable by forming cusps that elongate into finger-like jets. Fig. 5.1c illustrates how the cusps elongate to finger-like jets while the rim is further fed with liquid coming from the spreading droplet. The protruded finger-like jets elongate until surface tension forces cause a breakup into secondary drops.

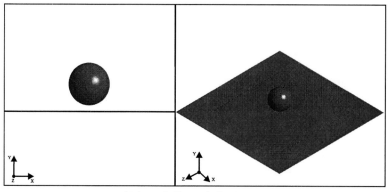

(a) $t = -0.15\ ms$ - before impact

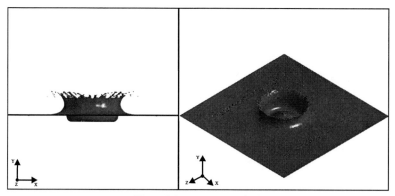

(b) $t = 1.4\ ms$ - prompt splash

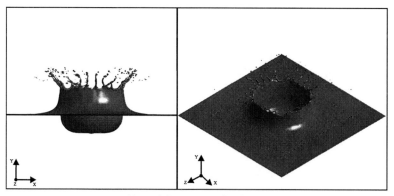

(c) $t = 2.5\ ms$ - crown breakup 1

Fig. 5.1: Fig. is continued in the next page

(a) $t = 5.6\ ms$- crown breakup 2

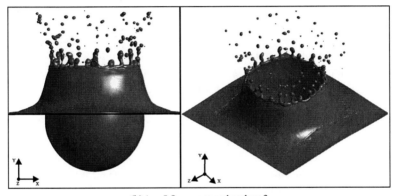

(b) $t = 8.2\ ms$- crown breakup 3

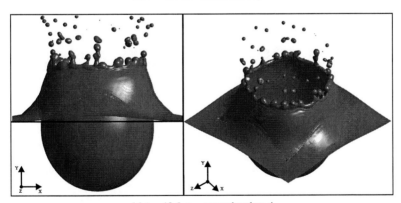

(c) $t = 12.6\ ms$- crown breakup 4

Fig. 5.2: Iso-surface of $\alpha = 0.5$ showing different stages of the crown's breakup during drop impact on a liquid layer (We=2209, Oh=0.0031)

The build-up of pressure caused by the enhanced surface tension forces at the edge of the elongated finger-like jets is well shown in Fig. 5.4. In this figure, the iso-surface of the volume fraction at $\alpha = 0.5$ is coloured by the calculated static pressure contours. During its expansion, the thickness of the crown and the sizes of the secondary drops increase in time (see the Figs 5.1b-5.2c). The breakup process lasts until the crown starts to retract and falls back. At this time, large secondary drops are separated that carry the major part of the ejected secondary drop mass (see in Fig. 5.2c).

5.2   Analysis of the Instability driving the Disintegration of the Crown

The instability responsible for the crown's breakup has been the subject of many experimental and numerical studies. An agreement on the underlying mechanism is, however, (still) not found in the scientific literature. The most common stated mechanism sees the Rayleigh-Taylor (RT) and the Rayleigh-Plateau (RP) instability dominating the crown's breakup. Both theories follow the physical idea that the torus-like shape accumulating at the rim of the lamella behaves like a cylinder of fluid subject to surface tension forces. The RT instability is additionally amplified by the rim deceleration. In the present section, a comparison of the predicted instability using the self-perturbing 3D-VOF-AMR technique with the most common theories is attempted. The reference impingement conditions are selected as such to keep the Ohnesorge number very low, specifically with $We = 2209$, $Oh = 0.0031$, and $Fr = 343$. With this, viscous (damping) effects influencing the wave length selection can be neglected. In order to enable comparison with the RT and RP theory, two assumptions are needed; (i) the crown can be described as a cylindrical sheet and is (ii) detached from the lamella. This sheet is subject to surface tension and continuously perturbed by a broad band of disturbances.

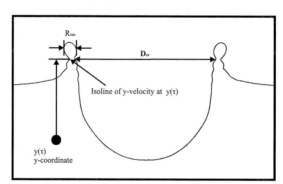

Fig. 5.3: Different parameters describing the crown's rim

Following the linear stability theory, when the fluid sheet is subject to a broad band of initial perturbations and the wavelength of the instability is larger than a characteristic value, it becomes unstable and may break up. Fig. 5.4 depicts the iso-surface of the volume fraction at $\alpha = 0.5$ which is coloured by the absolute velocity contours.

The y-coordinate of each point located on the isoline of the velocity near the rim of the lamella was extracted for four different time instances and plotted versus the circumference of the lamella in this point in Fig. 5.5. Here, the fluctuation of each curve indicates the small capillary waves perturbing the sheet continuously and the broad band of perturbations. Moreover, their amplitude appears to rise with time. The mean values of the y-coordinate increase in time since the iso-lines were extracted at positions near the perturbed rim, which increases in height over the observed time period. The circumference of the lamella also increases each time instances, τ, therefore the length of the curves rises, too.

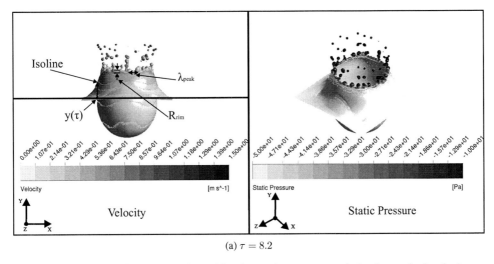

(a) $\tau = 8.2$

Fig. 5.4: Iso-surface of $\alpha = 0.5$ coloured by the static pressure and absolute velocity during a crown's breakup: We=2209, Oh=0.0031, and Fr=343

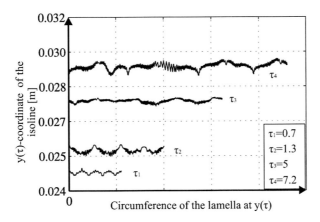

Fig. 5.5: Capillary corrugations perturbing the crown's rim (We=2209, Oh=0.0031 and Fr=343)

For each predicted crown disintegration stage during the complete impingement evolution, a number of quantities describing the crown's motion (torus) and shape are extracted from the self-perturbing 3D-VOF-AMR simulation. The diameter of the cylindrical sheet and its length (circumference) are therefore plotted versus the non-dimensional time τ (see Fig. 5.6). Figs 5.6a and 5.6b show clearly that the thickness of the torus-like shape at the rim increases together with its rim diameter (length) with time. Moreover, in the Figs 5.6c and 5.6d, the temporal evolution of the rim trajectory and the deceleration are further more depicted. The latter two quantities are required for comparison with the RT instability. Here, the strong deceleration of the rim in the early stages of the impingement is clearly visible. Hence, the geometrical shape, the trajectory and the motion of the torus-like shape are continuously changing. This makes a pure analytical prediction of the crown's breakup very complex and nearly impossible since the wave length of the fastest growing mode is also continuously reallocated. With the calculated flow field at hand, however, the crown's disintegration stages are accurately identified. Hence, the nature of instability driving the breakup in this stages can be compared to the RT and RP theories.

When the RP or the RT instability may dominate the crown's breakup, the wave length of the fastest growing mode and the number of cusps forming at the crown in each disintegration stage should correspond to the theory. In Figs 5.6f and 5.6e, a comparison of the calculated 3D-VOF-AMR number of cusps n_{cusps} and the peak wave length λ_{peak} in time with the associated predictions of the RT and RP theory is shown. The comparison proves clearly the dominance of the RT instability in the early stages of the impingement, at $\tau < 1.5$, whereas the RP instability as the driver in the later stages of the crown's breakup, at $\tau > 1.5$. For $\tau < 1.5$, the calculated peak wave length of the 3D-VOF-AMR is much larger than the one predicted by the RP theory but in excellent agreement with the RT theory. In the later stages of the impingement, for $\tau > 1.5$, the calculated peak wave length is significantly smaller than the one predicted by the RT theory while providing very good comparison with the RP theory. The strong deceleration of the crown in the early stages of the impingement evolution sees hence the RT instability driving the breakup. When the rim deceleration becomes small, breakup is mainly activated by the pure capillary driven RP instability. This interesting outcome may also explain why several authors stated either the RT or the RP instability as the driver (see, e.g. Rieber and Frohn (1999) and Zhang et al. (2010) or Krechetnikov and Homsy (2009)). The present outcome of the instability analysis corresponds well to the recent investigations of Roisman et al. (2006), Agbaglah et al. (2013), and Agbaglah and Deegan (2014)) who analytically proved the combined dominance of the RT and RP instability in the crown splash problem. Another important quantity significantly affecting the wavelength selection during the crown's breakup, and neglected in the present section, is the liquid viscosity. As will be shown later, higher viscous drop impingements appear to eliminate the disturbances in the early stage of the impingement; hence to cancel the Rayleigh-Taylor instability.

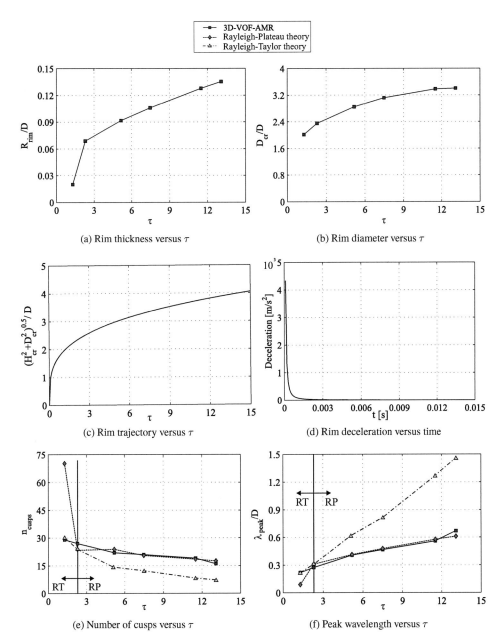

Fig. 5.6: Typical parameter describing the crown's instability during single drop impact onto liquid surfaces (We=2209, Oh=0.0031 and Fr=343)

In conclusion, the analysis of the crown's instability has shown for the underlying impingement conditions that a combined mechanism of the RT and the RP instability drives the breakup during single drop impact onto liquid surfaces. The fact that the crown in the drop splash problem may get unstable even without releasing secondary drops, and the continuous change in the shape and motion, makes an analytical prediction of the secondary drop characteristics nearly impossible. Hence, with respect to the drop-film interaction model applicable in bearing chamber simulation methods, correlations are needed that predict the final splashing products by the pre-impingement parameter of influence. This procedure requires a systematic parametric study that investigates the influence of various isolated parameters on the crown's position, secondary drop characteristics, and ejected mass fraction. This is accomplished in the next sections. All of these splashing products are needed for the formulation of a drop-film interaction model.

5.3   Evaluation Procedure for the Following Parametric Study

In the next sections, the analysis of several influencing impinging drop parameters, namely the Weber number, the Froude number, the Ohnesorge number, and the impingement angle, will be discussed. Since there is a systematic description procedure adopted for each parameter of influence, a short description of the evaluation procedure is attempted first. At the beginning of each section, the test conditions for the analysis of the splashing products (3D-VOF-AMR) and the lamella evolution (2D-VOF-AMR) are listed in tables. The different parameters of influence are then discussed by analysing the secondary drop generation qualitatively using the visualization of the iso-surface of the fluid volume fraction at the value $\alpha = 0.5$. Then, the quantitative analysis of the splashing products is carried out with the two-dimensional crown expansion during drop impingement. A detailed understanding of the crown expansion is needed in order to approximate the release position of secondary droplets later in the drop-film interaction model. Here, it was proven in Chapter 2 and in the experimental validation chapter that a two-dimensional axisymmetric method captures these dynamics sufficiently well. This method was therefore favoured since it requires less computational time than the 3D-VOF-AMR and thereby allows a larger number of parameter variations. Hence, all the numerical simulations analysing the effects on the crown/lamella evolution are conducted using the 2D-VOF-AMR method. The concluding part of each section determines each impinging effect on the most relevant impingement outcome, the secondary droplet characteristics, and the ejected mass fraction, using the 3D-VOF-AMR technique. Hence, in the next sections, an impinging parameter study is presented and discussed.

5.4 Effect of the Impinging Weber number

Drops within sprays in the oil systems of aero engines occur and interact with wall-films with different terminal impingement velocities and surface tension forces, depending on the operating conditions of the aero-engine. The effect of both quantities is usually well represented by the Weber number.

case	We	Fr	Oh	focus
1	389	28542	0.016	Lamella evolution
2	792	58249	0.016	Lamella evolution
3	1340	98441	0.016	Lamella evolution
4	1787	131061	0.016	Lamella evolution
5	353	295	0.0022	Break-up
6	520	455	0.0022	Break-up
7	683	642	0.0018	Break-up
8	755	343	0.0018	Break-up
9	1226	343	0.0023	Break-up
10	2209	343	0.0031	Break-up
11	792	64956	0.036	Break-up
12	1320	108373	0.036	Break-up
13	1850	151833	0.036	Break-up
14	2376	195072	0.036	Break-up

Table 5.1: Test conditions for the effect of Weber number

In this section, the effect of the impinging Weber number on the crown expansion and secondary drop generation is evaluated by selecting typical impingement scenarios relevant to these bearing chambers, where the terminal velocity and the surface tension coefficient are varied in isolation. All remaining parameters of influence, except the Froude number, are kept nearly constant for each set of impingements (cases 1-4, cases 5-10, and cases 11-14). As will be shown later, the effect of the Froude number on the impingement evolution requires a change of the Froude number of several order of magnitudes. In this section, the effect of the Froude number for each set of simulations is expected to be low. The test conditions are listed in Table 5.1.

Qualitative Description

The qualitative description of the influence of the Weber number on the splashing products is conducted using cases 8-10 in Table 5.1. Here, the individual figures of Fig. 5.7 show the effect of the Weber number for three different crown disintegration stages, namely $\tau = 1.3$, $\tau = 2.3$, and $\tau = 5.2$. An increase of the impinging Weber number results in an enhanced dominance of the kinetic energy over the surface tension energy. This, in turn, may indicate that fewer forces are available during the impingement to act against the additional surface generation, thereby making the release of secondary drops easier. By comparing the Figs 5.7a-5.7c, 5.7d-5.7f, and

5.7g-5.7i, it is evident that an augmentation of the Weber number causes larger crown heights and less liquid accumulation at the crown's rim. Moreover, the reduced dominance of the surface tension forces appears also to amplify the capillary disturbances and perturbations at the crown's rim. This causes an enhanced number and length of the protruded finger-like jets together with more ejected secondary droplets.

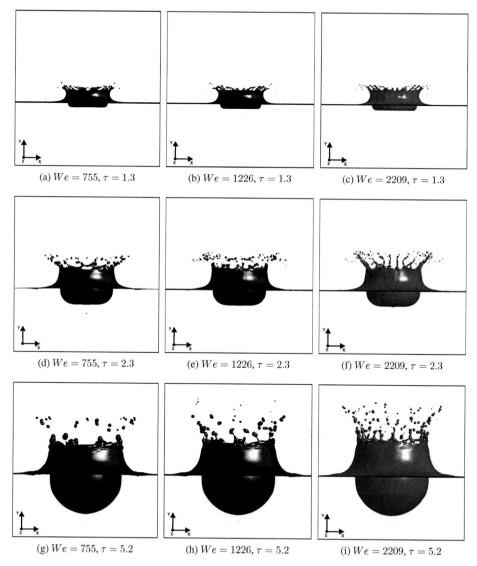

(a) $We = 755, \tau = 1.3$ (b) $We = 1226, \tau = 1.3$ (c) $We = 2209, \tau = 1.3$

(d) $We = 755, \tau = 2.3$ (e) $We = 1226, \tau = 2.3$ (f) $We = 2209, \tau = 2.3$

(g) $We = 755, \tau = 5.2$ (h) $We = 1226, \tau = 5.2$ (i) $We = 2209, \tau = 5.2$

Fig. 5.7: Qualitative depiction of the Weber number influence on the crown's breakup process during single drop impingement onto liquid surfaces

Position of the crown

By analysing the effect of the Weber number on the complete evolution of the crown expansion with the help of the test cases 1-4 specified in Table 5.1, the qualitative descriptions in the previous sections can be confirmed. Here, Fig. 5.8a depicts the lamella rim height and diameter versus impingement time. An increase of the Weber number from $We = 389$ to $We = 792$, leads to a change of the velocity of the expansion of the lamella rim and height by a factor of two in the early stage of the impingement. However, in this early stage of the impingement, the lamella rim diameter in Fig. 5.8b remains unaffected by a change of the Weber number.

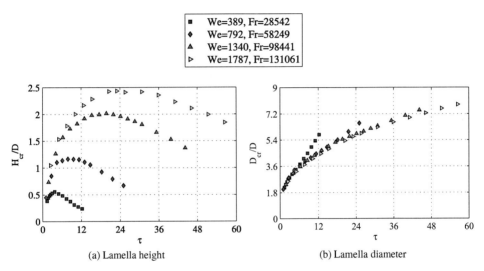

(a) Lamella height   (b) Lamella diameter

Fig. 5.8: Effect of Weber number on the evolution of the lamella rim: $Oh = 0.0161$, $D = 0.175 \ mm$

The expansion of the lamella rim continues until the completion of the conversion of the kinetic energy into surface tension energy, potential energy, and dissipated energy, and the maximum lamella rim height is approached. A typical maximum lamella rim height, for instance, results for the case with the largest Weber number $We = 1787$ with $H_{cr}/D \approx 2.5$, respectively. From here on, the lamella rim stops growing in height and starts falling back. A step-wise reduction of the Weber number leads to a sooner falling back of the lamella rim. This behaviour is well depicted in Fig. 5.8a. On the other hand, the evolution of the lamella rim diameter is slightly influenced by the impinging kinetic energy with lower non-dimensional lamella rim diameters with increased Weber numbers. This effect, however, gets weaker with increasing Weber number.

Secondary Drop Characteristics

A quantitative analysis of the arithmetic mean secondary drop sizes, velocities, and ejected mass fraction, is illustrated in Fig. 5.9.

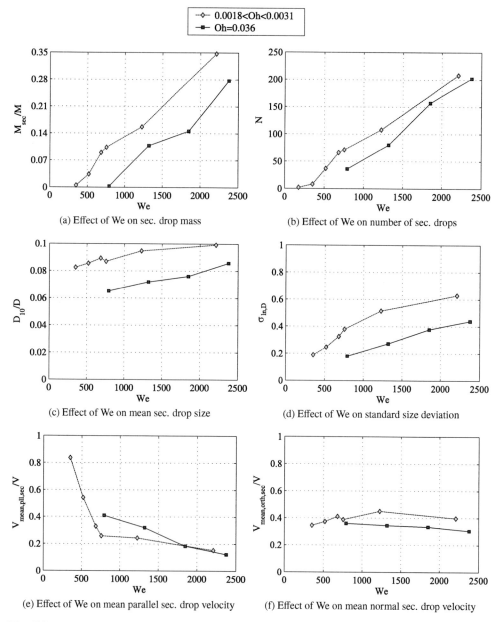

(a) Effect of We on sec. drop mass

(b) Effect of We on number of sec. drops

(c) Effect of We on mean sec. drop size

(d) Effect of We on standard size deviation

(e) Effect of We on mean parallel sec. drop velocity

(f) Effect of We on mean normal sec. drop velocity

Fig. 5.9: Effect of the Weber number on the characteristics and mass of the secondary drops during single drop impact onto deep pools

Here, a rise of the impinging Weber number from $We \approx 700$ to ≈ 1300 increases the ejected mass fraction and number of secondary drops by a factor of two. The increased number and mass of secondary drops with rising Weber number is however not only due to the production of more small secondary drops at the initial stage, but also significantly affected by the production of large secondary drops in the later disintegration stages. This is also confirmed by the linearly rise of the arithmetic mean secondary drop size D_{10} with the Weber number in Fig. 5.9c. The spread of the secondary drop size distribution, on the other hand, represented by the logarithmic standard deviation, augments with larger Weber number, too (see Fig. 5.9d). Figs 5.9e and 5.9f illustrate the effect of the Weber number on the mean parallel and orthogonal secondary drop velocity. The enhanced production of large secondary drops at the later stages of the disintegration with low absolute velocities and steep ejection angles show a significant decrease of the mean parallel secondary drop velocity with rising Weber number. The mean orthogonal secondary drop velocity, in contrast, is nearly constant with a weak reduction at very large Weber numbers, which may be assigned to the enhanced dissipation of kinetic energy compared to lower Weber numbers.

In conclusion, the Weber number has a significant effect on the lamella rim evolution during drop impact onto a semi-infinite liquid surface. The increase in impinging kinetic energy of droplets causes not only a larger temporal and spatial extension of the liquid film but also larger maximum lamella rim heights. During the evolution of the lamella rim, the effect of the impinging Weber number on the secondary drop characteristics and ejected mass fraction following the single drop impingement is significant. A rising Weber number has been shown to cause a nearly linear increase in the arithmetic mean secondary drop sizes with wider size distribution, and in the number and mass of the secondary drops. The secondary drops are released from the crown's rim with smaller absolute drop velocities and steeper ejection angles with increasing Weber number.

The attempt to apply the correlations of Samenfink et al. (1999), Tropea and Marengo (1999), and Okawa et al. (2006) for comparison with the secondary drop characteristics and ejected mass fraction presented in this chapter resulted in large variations, in particular for the large Oh numbers which are typically found in the oil system of aero-engines. The reason for the discrepancies may be explained, as identified by Cossali et al. (2005), by the narrow range of validity of these correlations, which does not cover the impingement conditions expected in the oil systems of aero-engines.

5.5   Effect of the Froude number Fr

The distributions of the sizes and angular velocities of the typical impinging sprays in an aero-engine oil system are known to vary over a wide range, namely between 5 μm and 2000 μm for the diameter and between 2 m/s and 20 m/s for the velocity (e.g. Birkenkaemper (1996), Glahn et al. (2002) and Glahn et al. (2003)). By simply replacing the gravitational with the centrifugal acceleration within the Froude number, there is a large variation in the Froude number denoted, too (see Chapter 2.3). The Froude number represents the ratio between the inertial and the gravitational/centrifugal forces. For a thorough investigation of the effect in isolation of the Froude number on the drop-to-film interaction, the only way to accomplish this is to change the gravitational or centrifugal acceleration during a single drop impingement. An experimental analysis of this effect is therefore very costly. Up to date, the only experimental investigation available in the literature is Kyriopoulos (2010), who investigated this effect on the liquid film hydrodynamics and spray cooling by parabolic flight routes. Here, the test rig was mounted on an aircraft. No other investigations of the effect of the gravitational/centrifugal forces on the drop impact dynamics have been found in the literature.

In this paragraph, the effect of the Froude number on the lamella rim evolution is studied by selecting a typical engine-relevant impingement scenario with $D = 0.175\ mm$ and $V = 15$ m/s. The gravitational/centrifugal acceleration is then changed step-wise from $g = 9.81\ m/s^2$ to $g = 2000\ m/s^2$. Hence, the range of Froude numbers investigated can be varied from the values relevant in practice up to large millimeter scale drop impacts. The impingement conditions used in this study are given in the cases 1-4 in Table 5.2.

The effect of the potential/centrifugal energy on the crown's breakup and the secondary drop characteristics is analysed a bit differently from the cases describing the lamella evolution. Since several authors have stated that the K-number is the crucial quantity affecting the spray generation during millimetre-sized drop impingement, it has been selected in this section as a scaling quantity ensuring similar non-dimensional impinging momentum. Hence, the isolated effect of the Froude number is separated by maintaining the K-number constant and varying only the Froude number. The range of K-numbers is chosen between $K = 4000$ and $K = 8565$ whereas the Froude number is varied between $Fr = 310$ and $Fr = 282928$. The significant variation of the Froude number emphasizes the strong change of the effect of the potential/centrifugal energy on the drop impact dynamics. The test cases of the 3D-VOF-AMR simulations used are the cases 5-17 in Table 5.2. The fluid properties were maintained constant in order to ensure the isolation of the parameter of influence in the present section.

case	K	Fr	Oh	focus
1	K=9340	642	0.016	Lamella evolution
2	K=9340	2571	0.016	Lamella evolution
3	K=9340	25712	0.016	Lamella evolution
4	K=9340	131061	0.016	Lamella evolution
5	K=4000	310	0.002	Break-up
7	K=6000	450	0.0218	Break-up
8	K=8565	343	0.0018	Break-up
9	K=4000	1409	0.0025	Break-up
10	K=6000	2037	0.0025	Break-up
11	K=8565	2908	0.0025	Break-up
12	K=4000	6247	0.0035	Break-up
13	K=6000	9366	0.0035	Break-up
14	K=8565	13392	0.0035	Break-up
15	K=4000	132112	0.0071	Break-up
16	K=6000	198068	0.0071	Break-up
17	K=8565	282928	0.0071	Break-up

Table 5.2: Test conditions for the effect of Froude number

Qualitative Description

The sequence in Fig. 5.10 shows the effect of three different impinging Froude numbers on the crown's breakup dynamics for a constant impinging K number of $K = 8565$ in three different breakup stages, $\tau = 1.3$, $\tau = 2.3$ and $\tau = 5.2$.

An enhancement of the impinging Froude number is equal to a reduction of the role of the gravitational forces that are acting against the crown's inertia. By comparing the qualitative depiction of the effect in the early stage of the impingement, at $\tau = 1.3$ and $\tau = 2.3$, in Figs 5.10a-5.10f, larger crown heights with increasing Froude number are clearly shown. The effect becomes more evident at $\tau = 5.2$ where not only do higher crown occur, but also a reduced crown thickness for higher Froude numbers. As a result of the reduced gravitational forces, the corrugations at the crown's rim appear also to be amplified with larger Froude numbers.

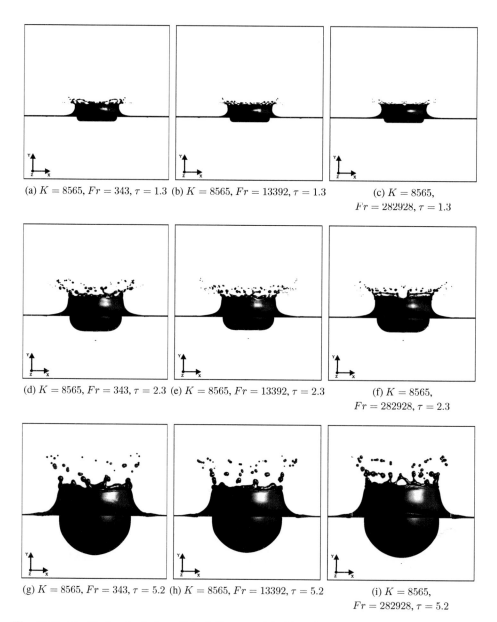

(a) $K = 8565$, $Fr = 343$, $\tau = 1.3$ (b) $K = 8565$, $Fr = 13392$, $\tau = 1.3$ (c) $K = 8565$,
$Fr = 282928$, $\tau = 1.3$

(d) $K = 8565$, $Fr = 343$, $\tau = 2.3$ (e) $K = 8565$, $Fr = 13392$, $\tau = 2.3$ (f) $K = 8565$,
$Fr = 282928$, $\tau = 2.3$

(g) $K = 8565$, $Fr = 343$, $\tau = 5.2$ (h) $K = 8565$, $Fr = 13392$, $\tau = 5.2$ (i) $K = 8565$,
$Fr = 282928$, $\tau = 5.2$

Fig. 5.10: Qualitative depiction of the influence of the Froude number on the crown's breakup process during single drop impingement onto liquid surfaces

Position of the Crown

The evolution of the height and diameter of the lamella rim is plotted versus τ in Figures 5.11a and 5.11b. The temporal evolution of the lamella rim height is influenced by the Froude number. However, it becomes visible only when the Froude numbers change from $Fr = 131061$ to $Fr = 2571$. Here, a change of the corona height from $H_{cr}/D \approx 2.5$ down to $H_{cr}/D \approx 2.3$ is shown. On the other hand, the maximum lamella rim height is reached for each increase in Froude number at later non-dimensional times. The lamella rim height expansion and falling back velocity remain nearly unaffected by a change of the Froude numbers. This is evidenced in Fig. 5.11a with the rising and falling slope of the curves. The evolution of the lamella rim diameter is not influenced by the Froude number and follows, according to the Weber number effect, a square-root dependence on τ (see Fig. 5.11b).

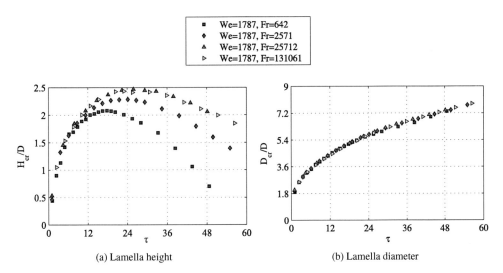

(a) Lamella height   (b) Lamella diameter

Fig. 5.11: Effect of Froude number on the evolution of the lamella rim: $We = 1787$, $Oh = 0.0161$, $D = 0.175\ mm$, $V = 15\ m/s$

Secondary Drop Characteristics

The arithmetic mean secondary drop diameter, velocity and ejected mass fraction resulting from this effect can be seen in Fig. 5.12. The amplification of the secondary drop generation identified in the qualitative description of this phenomenon is depicted by the plots of the ejected mass fraction and the number of secondary drops in Figs 5.12a and 5.12b. Here, an increase in the impinging Froude number from $Fr = 642$ to $Fr = 282928$ for a constant $K = 8565$ results in an enhanced secondary drop mass, from $M_{sec}/M \approx 0.085$ to $M_{sec}/M \approx 0.24$, and number of secondary drops, from $N = 65$ to $N = 122$. However, the effect decreases significantly with lower values of K. For $K = 4000$, it nearly disappears.

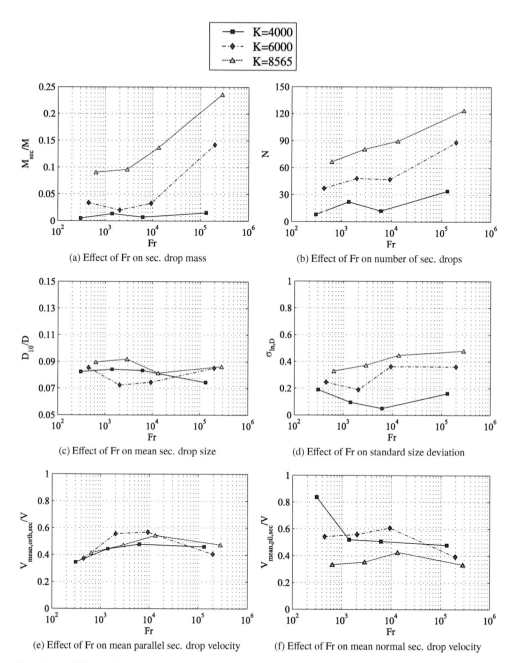

(a) Effect of Fr on sec. drop mass

(b) Effect of Fr on number of sec. drops

(c) Effect of Fr on mean sec. drop size

(d) Effect of Fr on standard size deviation

(e) Effect of Fr on mean parallel sec. drop velocity

(f) Effect of Fr on mean normal sec. drop velocity

Fig. 5.12: Effect of Froude number on characteristics and mass of secondary drops during single drop impact onto liquid surfaces

The arithmetic mean diameter in Fig. 5.12c, however, appears not to be decisively affected by the impinging Froude number. This implies that the additional secondary drops produced with larger impinging Froude number are equally released in small and large sizes. An equal additional production of small and large secondary drops with a nearly constant arithmetic mean secondary drop diameter is only possible by increasing the spread of the size distribution, which is correctly represented by the logarithmic standard deviations in Fig. 5.12d. Here, a clear increase in the logarithmic standard deviation with larger Froude number is observable. The parallel and orthogonal secondary drop velocities appear mostly independent of the Froude number for a constant level of K numbers since no clear trend can be derived from the graphs in Figs 5.12e and 5.12f.

In conclusion, in this paragraph a thorough analysis of the isolated influence of the Froude number on the lamella rim evolution and the secondary drop generation was conducted. With respect to the crown evolution, the Froude number played a role. However, it is only noticeable for Froude numbers smaller than $Fr = 25712$ whereas the lamella rim diameter proved to be independent of the Froude number. Moreover, the effect of the Froude number on the secondary drop generation revealed some very interesting observations, which have not, to date, been addressed in the literature. The reduction of the role of the gravitational/centrifugal forces acting at the crown's rim, by increasing the Froude number and maintaining the non-dimensional momentum K constant, enhances the secondary droplet generation at the crown's rim. Not only were larger ejected mass fractions identified, but also a considerably wider spread of the size distribution. This was identified for K numbers larger than $K > 4000$. However, the arithmetic mean secondary drop sizes together with their parallel and orthogonal ejection velocities appeared to be independent of the Froude number. Hence, a clear difference between drop impingement scenarios with low and high Froude numbers is identified at this stage. The latter directly reveals the shortcoming of millimetre sized impinging drops as a substitute for the technical relevant submillimeter impingement conditions. All drop-film interaction models taking only the non-dimensional momentum (i.e. the K number) into account are missing the important role of the gravitational/centrifugal forces during the impingement dynamics (see, e.g. Mundo et al. (1995), Samenfink et al. (1999), Tropea and Marengo (1999), Okawa et al. (2006), and Gomaa and Weigand (2012)). An extrapolation of those models to different ranges of impinging Froude numbers is therefore questionable. With the correlations developed later in this thesis based on each investigated influence, the effect of the potential/centrifugal energy will be directly included through the additional superposition of the Froude number. Hence, a considerable improvement of the actual secondary droplet predictions in a bearing chamber will be achieved. In bearing chambers, the effect of the gravitational/centrifugal acceleration on the secondary droplet generation is also very important. It is significantly determined by the rotational speed of the shaft since both the impinging droplets and the liquid wall films undergo a significant centrifugal acceleration in addition to the gravity.

5.6   Effect of the Impinging Ohnesorge Number

Depending on the operating conditions of the aero-engine, the oil temperature within the oil system can vary over a wide range, i.e. between $-20\ °C$ and $250\ °C$. For the drop-to-film interaction in bearing chambers, this causes, in turn, a significant change of the fluid's properties: in particular, the oil viscosity varies over several orders of magnitude. In order to investigate the effect of the oil viscosity on the drop impact dynamics, the Ohnesorge number proves to be a good scaling quantity. It represents the ratio between the viscous forces and the surface tension forces.

The test conditions for the investigation of the Ohnesorge number effect on the crown evolution and secondary drop generation are listed in Table 5.3. For the analysis of the crown evolution, three different fluids were defined: the liquid viscosity is varied between $\mu_l = 0.00102\ \frac{kg}{ms}$ and $\mu_l = 0.015\ \frac{kg}{ms}$. The selected range of liquid viscosity corresponds well to the conditions found in the oil systems of aero-engines. The effect of the liquid viscosity on the secondary drop generation is analysed using three different magnitudes of the Weber number. Since very large impinging Weber numbers are needed for large Ohnesorge numbers to ensure a crown break-up, only the test cases 4 and 5 (with very large impinging We) are used for the qualitative description of the crown disintegration. The remaining test cases for break-up are then used for the evaluation of the secondary drop characteristics and ejected mass fraction. All remaining parameters of influence are kept constant.

case	We	Fr	Oh	focus
1	1787	131061	0.016	Lamella evolution
2	1787	131061	0.08	Lamella evolution
3	1787	131061	0.24	Lamella evolution
4	2209	343	0.0031	Break-up
5	2209	343	0.047	Break-up
6	755	343	0.0018	Break-up
7	879	343	0.011	Break-up
8	879	343	0.026	Break-up

Table 5.3: Test conditions for the effect of Ohnesorge number

Qualitative Description

A qualitative illustration of the effect of the liquid viscosity on the crown's disintegration is shown in Fig. 5.13 for three different time instances using test cases 4 and 5 in Table 5.3. In the early stage of the splashing impingement, at $\tau = 1.3$, an increase of the Ohnesorge number from $Oh = 0.0031$ to $Oh = 0.047$ clearly depicts the additional damping of the crown's corrugations since no secondary droplets were generated in this stage of the process for the larger Ohnesorge number.

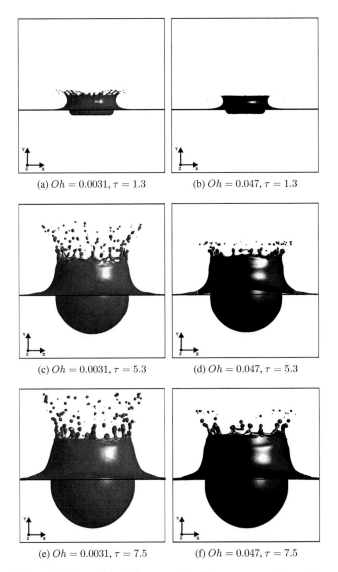

(a) $Oh = 0.0031$, $\tau = 1.3$ (b) $Oh = 0.047$, $\tau = 1.3$

(c) $Oh = 0.0031$, $\tau = 5.3$ (d) $Oh = 0.047$, $\tau = 5.3$

(e) $Oh = 0.0031$, $\tau = 7.5$ (f) $Oh = 0.047$, $\tau = 7.5$

Fig. 5.13: Qualitative depiction of the influence of the Ohnesorge number on the crown's breakup process during single drop impingement onto liquid surfaces: $We \approx 2300$ and $H^* >> 6$

However, in the further evolution of the impingement, at $\tau = 5.3$ and $\tau = 7.5$, corrugations take place at the crown's rim for both cases with different wavelength but similar amplitudes. The stronger damping effect for the impingement $Oh = 0.047$ shows a smaller number of finger-like jets protruding from the crown's rim. Due to the correspondence of the crown's height, diameter, and thickness for both impingements, only the wavelengths and amplitude of the perturbations travelling along the lamella are damped by an increase of the Ohnesorge number. This in turn leads to the identified change of the crown's corrugations. The damping of the latter causes not only in the initial stage, but also in the intermediate and end stage of the impingement a significant reduction of number and mass of secondary drops (see Figs 5.13c-5.13d and 5.13e-5.13f). However, the released secondary drop sizes and ejection angles from the crown are comparable for both impingements. This observations agree well with the one identified by Zhang et al. (2010) since nearly no micro droplets produced from the prompt splashing regime were found for impingements with low Re and high We numbers.

Position of the Crown

A description of the complete evolution of the lamella is depicted using the temporal evolution curves in Fig. 5.14 based on the test cases 1-3 in Table 5.3. The effect of increasing viscosity becomes significant in the lamella rim expansion regime, where smaller rim heights exist due to the enhanced damping of the viscous forces. However, this effect is only visible with very large Ohnesorge numbers, specifically from $Oh = 0.016$ to $Oh = 0.24$.

When the lamella rim reaches its maximum vertical extension, a noticeable influence of μ_l exists that is reflected again by a reduction of the lamella rim height from $H_{cr}/D = 2.5$ to $H_{cr}/D = 1.5$. The maximum lamella heights are reached sooner for larger Ohnesorge numbers. Interestingly, the initiation of the lamella rim retraction, and the lamella rim retraction velocity is similar for all the investigated cases (see Fig. 5.14a). With respect to the lamella rim diameter, no noticeable effect can be identified in Fig. 5.14b. Hence, in contrast to several reports in the literature, here we find that only very large changes in the Ohnesorge number have a noticeable effect.

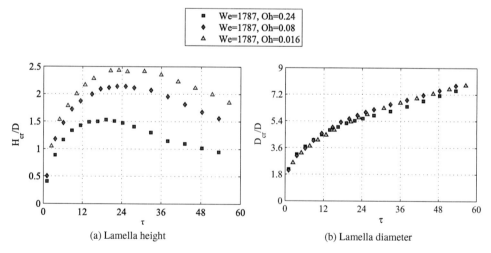

(a) Lamella height   (b) Lamella diameter

Fig. 5.14: Effect of Ohnesorge number on the evolution of the lamella rim during single drop impingement onto a deep liquid pool: $We = 1787$, $Fr = 131061$, $D = 0.175\ mm$, $V = 15\ m/s$

Secondary Droplet Characteristics

An increase in the impinging Ohnesorge number leads to an enhanced dissipation of the capillary disturbances and thereby it causes fewer corrugations at the crown's rim. This damping effect is quantitatively shown by the ejected mass fraction and the number of secondary drops in Figs 5.15a and 5.15b. Here, for instance, an increase in the Ohnesorge number from $Oh = 0.0031$ to $Oh = 0.047$ reduces the number and mass by approximately a factor of two. On the other hand, the arithmetic mean secondary drop sizes in Fig. 5.15c show a clear decrease only for the cases of the larger Weber numbers (We=2300, cases 4 and 5) while approaching a constant level for the other test cases with lower Weber numbers (We=800, cases 6, 7, and 8). In contrast, the spread of the size distribution with the logarithmic standard deviation in Fig. 5.15d is only reduced for the test cases with smaller Weber numbers , $We = 800$, while having a nearly constant trend for the test cases 4 and 5 with large Weber numbers. Only for the test cases with lower Weber numbers (We=800, cases 6, 7, and 8), does the increased damping of the crown's breakup lead to a reduction of the release of large secondary drops in the later stages of the disintegration process; this confirms the smaller arithmetic mean sizes and standard deviation. The parallel and orthogonal release velocities of the secondary drops are depicted in the Figs 5.9e-5.9f. Here, a considerable increase in the velocity component parallel to the liquid surface can be seen. On the other hand, the effect of the Ohnesorge number on the release velocity orthogonal to the undisturbed liquid surface is very weak (see Fig. 5.9f).

In summary, it has been demonstrated that for kinematic impact conditions close to those found in the oil systems of aero-engines, the viscosity of the liquid has a decisive impact on the complete lamella rim evolution only for very large changes in values of the Ohnesorge number. This range of Ohnesorge numbers does occur in bearing chambers. Therefore, it is prerequisite for the later modelling of this interaction. An increase in the liquid viscosity results in smaller maximum lamella rim heights and reduced time needed to reach the maximum lamella height. With respect to the secondary drop generation, the augmentation of the Ohnesorge number acts via damping of the capillary forces by reducing the number of finger-like jets at the crown's rim and thereby the ejected mass and number of secondary drops. The mean secondary drop sizes and spread of the distributions diminished slightly with this increased damping. The secondary drop velocities appeared to be affected only parallel to the liquid surface, while being mostly independent of Oh in the normal release direction. With the results obtained in this section, it has been clearly shown that the probability of extensive secondary drop generation in bearing chambers is only possible in warm engine operating conditions where the oil viscosity and the Ohnesorge number are very low. For these conditions, depending on the impinging kinetic energy, a damping of the forces leading to the crown's breakup is very low, as was demonstrated in the qualitative description. However, when the oil is cool in cold start engine cycles and low ambient temperatures, the probability of occurrence of secondary drops due to drop-to-film interaction in bearing chambers is very low.

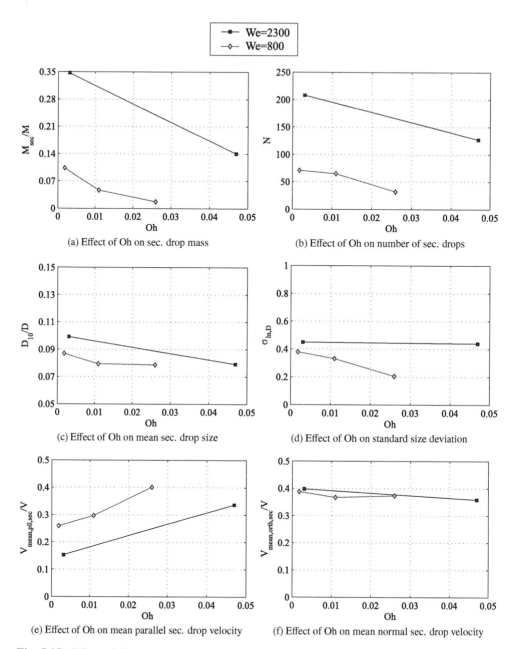

Fig. 5.15: Effect of Ohnesorge number on the characteristics and mass of the secondary drops during single drop impact onto liquid surfaces

5.7 Effect of the Impingement Angle α

Within sprays in aero-engine oil-systems, the impingement of the oil droplets on the wall films is not always normal but frequently at an oblique angle. This is primarily caused by the strong interaction of the droplets and the rotating air flow driven by the engine shafts. Small droplets within the sprays partly follow the path of the air flow, due to their low inertia. Hence, the resulting impingement angles between the droplet and the wall-film are much flatter than would be those for large drops. Despite the practical relevance of this, to date the number of publications characterizing and modelling oblique impingements is rather small.

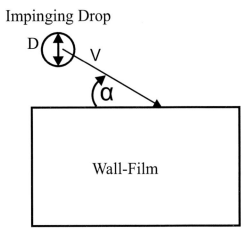

Fig. 5.16: Angle description for oblique impingement simulations

No numerical methods for the description of the secondary drop characteristics and the ejected mass fraction in oblique impingement dynamics could be found in the literature. This section offers therefore one of the first detailed analyses of the effect of the impingement angle on the lamella evolution and secondary drop generation with the help of numerical modelling. It will contribute not only to an improved modelling of practical drop impingement scenarios, but also enhance the fundamental understanding of drop impact dynamics in general. The effect of the impingement angle on the lamella rim evolution will be studied using a reference drop impingement with the following parameters: We=541, Oh=0.0088 and Fr=131061. Secondary drop generation under different drop impingement angles will be analysed for two different magnitudes of impinging Weber numbers, $We = 1320$ and $We = 1850$, respectively. The impingement angle is varied between $\alpha = 90°$ and $\alpha = 30°$. The sketch in Fig. 5.16 shows how the impingement angle is defined. The test cases are listed in Table 5.4.

case	We	Fr	Oh	Impingement angle	focus
1	541	131061	0.0088	90	Lamella evolution
2	541	131061	0.0088	60	Lamella evolution
3	541	131061	0.0088	45	Lamella evolution
4	541	131061	0.0088	30	Lamella evolution
5	1320	108373	0.036	30	Break-up
6	1320	108373	0.036	45	Break-up
7	1320	108373	0.036	60	Break-up
8	1320	108373	0.036	90	Break-up
9	1850	151833	0.036	30	Break-up
10	1850	195072	0.036	45	Break-up
11	1850	195072	0.036	60	Break-up
12	1850	195072	0.036	90	Break-up

Table 5.4: Test conditions for the effect of impingement angle

Qualitative Description

The effect of three different impact angles on the crown's breakup for an impinging Weber number of $We = 1850$ is depicted in a side and top view in Figs 5.17 and 5.18 at different impingement times. For a better impression of the crown's breakup dynamics, iso-surfaces of the volume fraction at $\alpha = 0.5$ in the top view are additionally coloured by the y-coordinate. The qualitative visualizations in the Figs 5.17 and 5.18 clearly reveal a change in the impingement dynamics with a variation of the incident angle. The drop impact dynamics becomes more and more asymmetrical with respect to the crown's shape around the impingement point at lower angles of incidence. In the initial stage of the impingement evolution, at $\tau = 1.8$, non-normal impingement causes a crown only downstream from the impingement point. The lamella rim in the upstream direction is unperturbed and does not release any secondary drops for all oblique impingement angles. Production of secondary drops during oblique impingement is hence concentrated mainly downstream. With the impact angle getting flatter, the height of the crown reduces, too. The geometry of the lamella departs from its sperical shape and elongates more and more to an ellipsoid. For the smallest impact angle, at $\alpha = 30°$, the crown even re-joins the liquid film and produces a much lower number of secondary drops. This phenomenon may be explained by the occurrence of enhanced dissipation of kinetic energy and axial momentum transfer to the liquid film taking place due to mixing and sliding of the drop into the liquid film. The axial direction refers here to the component parallel to the liquid surface. The counter acting liquid shear force decelerates the sliding and penetrating droplet and allows the gravitational and surface tension forces to pull-down the lamella towards the liquid film. Interestingly, at a later impingement stage, at $\tau = 4.7$, the downstream crown height of the impingement with an angle of $\alpha = 60°$ is evolving larger in height than the actual normal impingement. This interesting and unexpected behaviour may be explained by taking the energy balance into account. As a difference to the impingement angle of $\alpha = 30°$, a much lower fraction of dissipation of kinetic energy and axial momentum transfer to the liquid film takes place for the impingement with $\alpha = 60°$ since there is less sliding of the impinging drop.

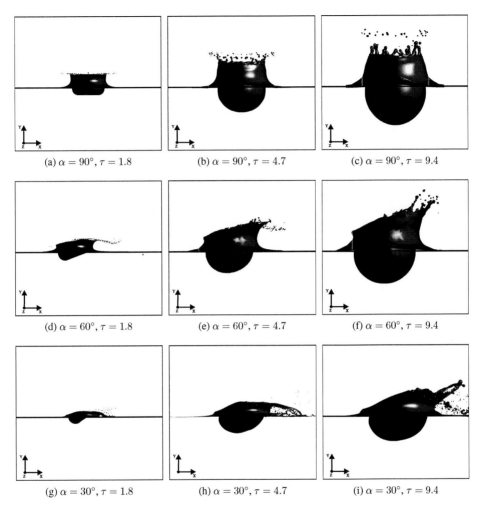

(a) $\alpha = 90°$, $\tau = 1.8$ (b) $\alpha = 90°$, $\tau = 4.7$ (c) $\alpha = 90°$, $\tau = 9.4$

(d) $\alpha = 60°$, $\tau = 1.8$ (e) $\alpha = 60°$, $\tau = 4.7$ (f) $\alpha = 60°$, $\tau = 9.4$

(g) $\alpha = 30°$, $\tau = 1.8$ (h) $\alpha = 30°$, $\tau = 4.7$ (i) $\alpha = 30°$, $\tau = 9.4$

Fig. 5.17: Qualitative depiction of the influence of the impact angle on the crown's breakup process during single drop impingement onto liquid surfaces from the side (We=1850, Oh=0.036, Fr=195072)

The less amount of dissipated energy can now additionally be invested and transferred in a more concentrated manner into the crown expansion and cause locally larger crown heights than the normal impingement. This behaviour is also maintained until the ending stage of the crown's breakup at $\tau = 9.4$. For the impingement with the flattest impact angle, only at the end stage of the impingement does a build-up of a lamella and crown occur. Large secondary drops are then released from the crown.

(a) $\alpha = 60, \tau = 1.8$ (b) $\alpha = 60, \tau = 4.7$ (c) $\alpha = 60, \tau = 9.4$

(d) $\alpha = 30, \tau = 1.8$ (e) $\alpha = 30, \tau = 4.7$ (f) $\alpha = 30, \tau = 9.4$

Fig. 5.18: Qualitative depiction of the influence of the impact angle on the crown's breakup process during single drop impingement onto liquid surfaces from the top (We=1850, Oh=0.036, Fr=195072)

They are also released for the $\alpha = 60°$ impingement although with considerably smaller sizes than for the flattest incident angle of $\alpha = 30°$ (see Figs 5.17f and 5.18c). The release of large secondary droplets at the end of the impingement stage was also confirmed in the experimental investigation of Okawa et al. (2006).

Crown Position

The effect of the impingement angle is depicted in Fig. 5.19 with the temporal evolution of the lamella rim. A decrease of the impingement angle from $\alpha = 90°$ to $\alpha = 60°$ results in a large increase of the lamella height. The much lower fraction of axial impingement energy is not sufficient to form a large lamella sliding in the positive axial direction above the liquid film. Therefore, the entire fraction of impinging kinetic energy is mainly transferred into a vertical expansion of the lamella. For the two lower impingement angles, $\alpha = 45°$ and $\alpha = 30°$, the axial fraction of the kinetic energy is large enough to form a lamella evolving in the axial direction above the liquid film. This behaviour can be seen by the much larger lamella rim diameter and smaller rim heights in Figs 5.19a and 5.19b. Here, the lamella rim diameter in oblique impingements is only an indicator to characterize the elongation of the spherical lamella to an ellipsoid. It is measured as the length of the ellipsoid in its symmetry plane. The more intensive sliding of the lamella over the liquid film causes also a larger rate of dissipation of kinetic energy. Correspondingly, the maximum lamella rim height increases first from $H_{cr}/D \approx 0.9$ to $H_{cr}/D \approx 1.35$ for $\alpha = 60$ and reduces abruptly to $H_{cr}/D \approx 0.65$ for $\alpha = 30$. The time to reach the maximum lamella rim heights is increasing with larger impingement angles. However, the lamella expansion velocity and retraction velocity, as depicted by the slope of their graphs, is not considerably affected by the impingement angle.

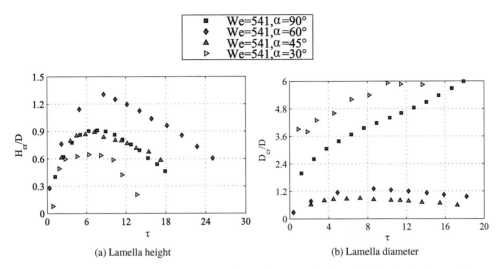

(a) Lamella height   (b) Lamella diameter

Fig. 5.19: Effect of impingement angle on the evolution of the lamella rim: $We = 542$, $Fr = 131061$, $Oh = 0.0088$

Secondary Drop Characteristics

According to the previously discussed influences, the quantitative analysis of the impact angle effect on the secondary drop characteristics and ejected mass fraction is depicted in Fig. 5.20. The quantitative analysis of the secondary drop characteristics and ejected mass fraction basically confirms the observations derived from the visualizations and the lamella evolution. The effect of the impingement angle shows first a strong increase in the secondary drop mass and number from impingement angles of $\alpha = 90°$ to $\alpha = 45°$. From here on, an enhanced dissipation and momentum transfer to the liquid film appears to dominate the impingement, and there occurs a significant reduction in the mass and number of secondary drops. For an impact with an impinging Weber number of $We = 1850$, a change of the impingement angle from $\alpha = 90°$ to $\alpha = 45°$ results in larger mass of the secondary drops released: by a factor of 3 compared to the normal impingement. For flatter angles, the ejected mass fraction is reduced to the same values as for the normal impingement. The arithmetic mean of the secondary drop diameters and the logarithmic standard deviation increase continually with flatter impingement angles. This augmentation is due to the larger number of large secondary drops released in the final stage of the crown's breakup. The changing slope of the linear decreasing trend of the arithmetic mean diameter for impingements with angles smaller than $\alpha = 45°$ can be attributed to the re-joining of the lamella with the liquid film in the initial stage of the impingement.

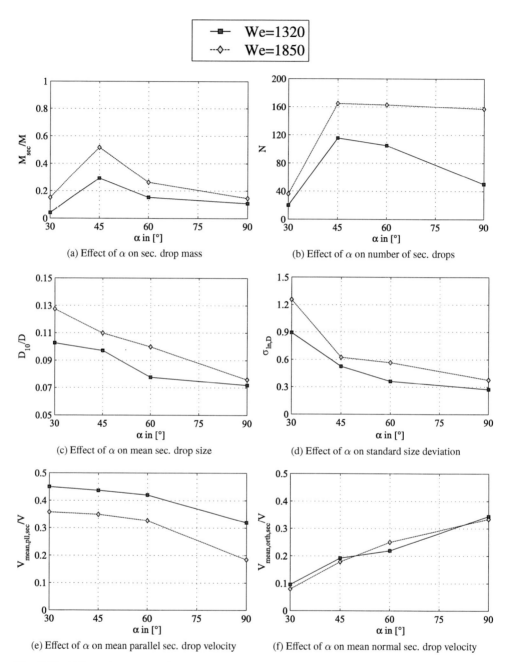

Fig. 5.20: Effect of impact angle on the characteristics and mass of the secondary drops during single drop impact onto liquid surfaces

This re-joining suppresses the production of small secondary drops, which in turn leads to a larger arithmetic mean of the secondary drop sizes compared to steeper impingement angles.

The concentration of the kinetic energy in the downstream direction at oblique impingements indicates that the release of secondary drops occurs with a higher mean parallel and a lower mean orthogonal velocity. The decrease of the velocity component of the secondary drops, normal to the undisturbed liquid surface, with flatter incident angles is significant compared to the parallel velocity component. This may be explained by the aforementioned increasing level of dissipation of kinetic energy for flatter angles.

From the results of the present subsection, it can be concluded that the effect of the impingement angle on lamella expansion, as quantitatively studied in this section for the first time, is considerable and needs to be taken into account in the modelling of spray impingements. The lamella rim expansion behaves in a much more complicated fashion and has non-linear trends. A reduction of the impingement angle was accompanied by an increase in the lamella rim height; the reduction is abrupt for impingement angles smaller than $\alpha < 60°$. The study of this effect on the characteristics of the secondary drops and on the ejected mass fraction proved the numerical method to be robust. Moreover, a very interesting insight into the oblique impingement dynamics was provided. A variation in the incident angle produced dramatic changes in the impingement dynamics, compared to the normal impingement. A larger number and mass of secondary drops resulted from a decrease of the impingement angle down to $\alpha = 45°$. This could also be confirmed for the arithmetic mean of their sizes and the logarithmic standard deviations. For smaller impact angles, the level of dissipation of kinetic energy together with momentum transfer to the liquid film increases due to the mixing and sliding of the spreading drop, leading in contrast to a marked reduction in the mass and number of secondary drops. The arithmetic mean drop size, however, continued to rise. With smaller impingement angles, the ejection of the secondary drops became more and more parallel to the liquid surface. Due to the increased level of dissipation of kinetic energy taking place for impingement angles smaller than $\alpha = 45°$, the absolute mean secondary drop velocity decreased more than the parallel velocity.

5.8   Concluding Discussion

The numerical results describing the influence of various parameters on the lamella rim expansion, and the secondary drop characteristics during single drop impingement onto deep liquid pools revealed several interesting phenomena. All of the knowledge produced in this chapter is not only of significant importance in the fundamental physics of drop impact dynamics but also relevant for different practical situations. With the help of the numerical database developed, correlations will be derived later in this thesis describing the splashing products in terms of the secondary drop characteristics and the ejected mass fraction. Hence, not only an improved understanding of the role of gravitational/centrifugal acceleration and fluid properties in drop-film interaction will be provided by the correlations, but also an enhanced prediction capability for oblique drop impingements. Unlike the available literature, these findings are valid for the range of impingement conditions found in aero-engine bearing chambers. In light of the results gained, hence, there is still the need to understand the role of the wall-effects in splashing drop impingements onto liquid surfaces, something which will be analysed and discussed in the next chapter.

6 Products of Splashing Drop Impingements with Wall Effects

In the last chapter, a thorough analysis of the secondary drop generation under various values of the impinging parameters was presented for drop impingement neglecting any effects coming from the wall. That is to say, the liquid film was considered to be of infinite height. In bearing chambers, the bounding walls are mostly covered by wall films which are very thin or thick. This depends on the operating point of the aero-engine (see Gorse et al. (2004), Kurz et al. (2012), and Hashmi (2012)). Hence, the droplets impinge on wall films with different heights. Therefore, the importance of understanding the role of the liquid film thickness in drop impingement dynamics is not only of fundamental relevance but also required in practical situations.

This chapter will analyse the effect of the liquid film height by relating the cavity evolution with the secondary drop generation for typical bearing chamber impingement conditions. It is divided into two main parts. In the first part, a detailed characterization of the cavity expansion dynamics and maximum cavity depth using deep pool impact is given using the axisymmetric 2D-VOF-AMR method. Several authors have already proved the accuracy of the 2D-VOF method for the description of the cavity penetration. The maximum cavity depth during deep pool impacts is needed for the later characterization of the influence of the wall film thickness on the secondary droplet generation for thin and shallow drop impacts. Here, if the maximum cavity depth for deep pool impingements is larger than the liquid film height of the corresponding thin or thick film impingements, a wall-effect is expected to take place. On the other hand, if it is smaller, then no wall effects need to be considered. Hence, the results of the last chapter assuming an infinite liquid film depth can then be used to describe the splashing products. A complete understanding of the latter is achieved by investigating step-by-step the most dominant effects on the cavity expansion. These are the kinetic energy (Weber number), the gravitation/centrifugation (Froude number), the fluid viscosity (Ohnesorge number), and the impingement angle. The second part of this chapter uses the knowledge gained in the first part to better characterize the effect of the liquid film on the secondary droplet generation by using the ratio $H^*/\Delta_{max,deep}$ and not only H^*. The numerical results of the previous Chapter 5, and the present Chapter, will build then the database from which correlations will be derived in the subsequent chapter.

6.1 Cavity Penetration during Deep Pool Drop Impact

6.1.1 Fundamental Description of the Cavity Evolution

A typical insight into the temporal evolution of the cavity for a submillimetre drop impact onto a liquid surface is depicted in Fig. 6.1. This is depicted by extracting the iso-surface of the liquid volume fraction at computational cell values of $\alpha = 0.5$ at different non-dimensional times τ from the calculated axisymmetric flow field. The visualizations are accompanied by the related pressure and velocity fields of the liquid film (see Fig. 6.4).

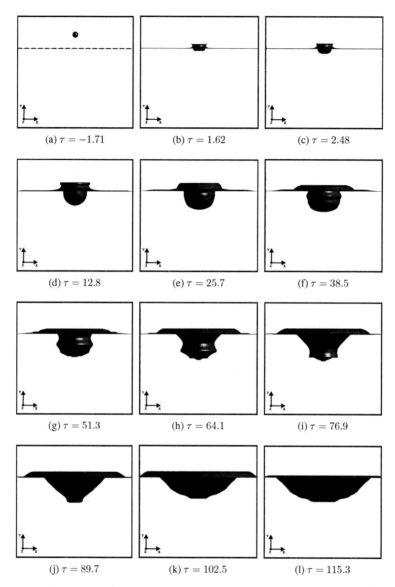

(a) $\tau = -1.71$ (b) $\tau = 1.62$ (c) $\tau = 2.48$

(d) $\tau = 12.8$ (e) $\tau = 25.7$ (f) $\tau = 38.5$

(g) $\tau = 51.3$ (h) $\tau = 64.1$ (i) $\tau = 76.9$

(j) $\tau = 89.7$ (k) $\tau = 102.5$ (l) $\tau = 115.3$

Fig. 6.1: Iso-surface of $\alpha = 0.5$ showing the temporal evolution of the cavity and lamella rim during drop impact on a liquid layer: We=541, Oh=0.0161, D=0.175 mm and V=15 m/s

In order to identify the differences in spatial evolution between the crown and the cavity, the temporal evolution curves of the cavity depth Δ, the lamella rim height H_{cr}/D, the cavity diameter D_{cav}/D and the lamella rim diameter D_{cr}/D are also shown (see Fig. 6.2) . The velocity field of the liquid film is depicted with the aid of randomly spaced arrows. The first contact of the impinging drop with the liquid surface occurs at $\tau = 0$.

After the formation of the ejecta and the lamella, a small cavity is formed. The cavity expands in depth and diameter while the lamella evolves radially outwards (see Figs 6.1b and 6.1c). Interestingly, the expansion velocities of both appear similar up to $\tau \approx 25$, as expressed in Fig. 6.2a.

The expansion of the lamella and the crater continues until the conversion of the impinging kinetic energy into potential energy, surface tension, and dissipated energy is entirely accomplished. However, the time to reach its maximum value appears much larger for the cavity depth evolution compared to the lamella height. Moreover, the maximum cavity penetration depth is much larger than the maximum lamella height (Figs. 6.1 and 6.2). The lamella reaches, for instance, its maximum height at around $\tau \approx 30$ whereas the maximum cavity depth is reached at $\tau \approx 250$. Moreover, the expansion into the liquid film is more than twice as great as the maximum lamella height. This implies also that a larger fraction of the impinging kinetic energy is available for the cavity than for the lamella expansion. The latter is in contrast to the reported behaviour of millimetre sized drop impingement, where the maximum lamella rim height and cavity depth are reached approximately at the same time. The shape of the cavity during its expansion is hemispherical in particular for drop impact scenarios, where the wall does not influence the cavity expansion (see Fig. 6.1). Once reaching their maximum height/depth, both the lamella and cavity start to retract. The lamella rim falls back rapidly due to gravitational forces and rejoins the liquid film.

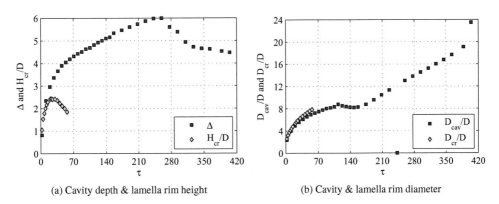

(a) Cavity depth & lamella rim height   (b) Cavity & lamella rim diameter

Fig. 6.2: Typical cavity and lamella rim expansion over time during a submillimeter drop impingement onto a deep liquid pool: $We = 1787$, $Oh = 0.0161$ and $Fr = 131061$

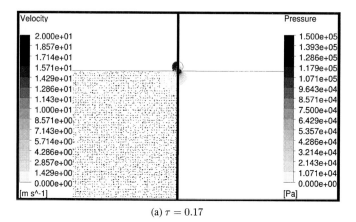

(a) $\tau = 0.17$

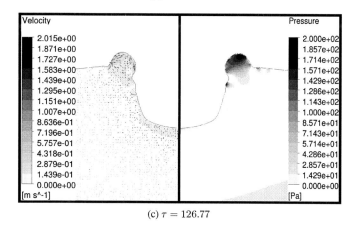

(b) $\tau = 11.05$

(c) $\tau = 126.77$

Fig. 6.3: Fig. is continued on the next page

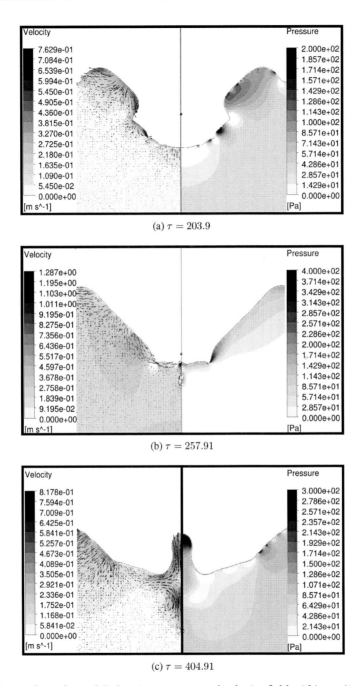

(a) $\tau = 203.9$

(b) $\tau = 257.91$

(c) $\tau = 404.91$

Fig. 6.4: Iso-surface of $\alpha = 0.5$ showing pressure and velocity field within cavity and lamella rim during drop impact on a liquid layer: We=541, Oh=0.0161, D=0.175 mm and V=15 m/s

While entering the liquid film, due to the larger curvature and gravitational acceleration forces taking place at the retracting rim, an elevated pressure occurs at the top of the rim that causes a pressure discontinuity between the rim and the liquid film (see Fig. 6.3c). The underlying pressure discontinuity initiates capillary waves, as experimentally investigated by Liow (2001) and Zhang et al. (2008), and numerically by Morton et al. (2000) and Berberović et al. (2009) (see Fig. 6.1f - 6.1h)). The initiation of the capillary waves is depicted by the pressure peaks in Fig. 6.3c.

For the underlying submillimeter drop impingement, however, these capillary waves appear much less pronounced in amplitude than for large millimetre sized droplets (see, e.g. Fig. 4.20e in Section 4.2.3). The capillary waves are transported further towards the bottom of the cavity by a difference in pressure occurring between the rim and the bottom of the crater (see Fig. 6.4a). The surface tension forces cause first a change in the cavity shape from a hemisphere to a conical one, by the downward motion of the capillary waves. Then, there is formed a central jet that may pinch off some drops (see Figs 6.1i-6.1j and 6.4b-6.4c).

In conclusion, it has been shown that for impingement conditions relevant to bearing chambers in particular, there are considerable differences between the cavity and crown dynamics during a drop impingement. These differences have not been reported in the literature, yet. With respect to the fundamental drop impingement dynamics, hence, there is a need for an improved understanding of the cavity evolution during single drop impingement. In addition to the fundamentals, the knowledge of the cavity penetration in this thesis is crucial for the characterization of the influence of the wall-film depth on the secondary droplet generation. In order to comply with both requirements, the next sections analyse the cavity evolution during single drop impingement by means of a parameter study.

6.1.2   Effect of the Froude Number on the Cavity Penetration

This paragraph presents the results of a numerical investigation of the effect of gravity/centrifugation on the evolution of the cavity. The study was conducted by selecting a typical engine-relevant impingement scenario: $D = 0.175\ mm$ and $V = 15\ m/s$ and subsequently varying step-wise the gravitational and centrifugal acceleration between $g = 9.81\ m/s^2$ and $g = 2000\ m/s^2$, respectively. The range of Froude numbers investigated, hence, could be varied from typical engine relevant values up to large millimetre scale drop impact conditions. The impingement conditions used in this study are the cases 1-4 listed in Table 5.2 of Chapter 5.

The qualitative depiction of the Froude number effect is shown in Fig. 6.5 for three different time instances, $\tau = 1.62$, $\tau = 12.8$ and at maximum cavity depth τ_{max}, respectively. Accordingly, the quantitative evolution of the cavity is plotted versus τ and compared to the recently developed analytical model of Bisighini et al. (2010) in Fig. 6.6. A comparison of the cavity evolution at the initial impingement stage $\tau = 1.62$ shows no clear differences for increasing gravitational/centrifugal acceleration from $Fr = 131061$ to $Fr = 642$ (see Fig. 6.5a-6.5c). This is also confirmed by Fig. 6.6, where the evolution of the cavity depth and the shape factor are corresponding for all the investigated test cases. The analytical model of Bisighini et al. (2010) appears also to resolve the physics of the cavity depth expansion phase up to $\tau \approx 35$. The influence

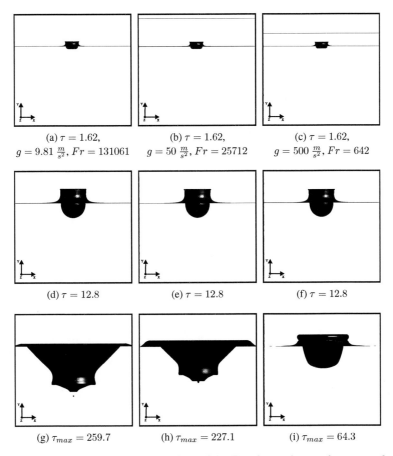

(a) $\tau = 1.62$,
$g = 9.81\ \frac{m}{s^2}, Fr = 131061$

(b) $\tau = 1.62$,
$g = 50\ \frac{m}{s^2}, Fr = 25712$

(c) $\tau = 1.62$,
$g = 500\ \frac{m}{s^2}, Fr = 642$

(d) $\tau = 12.8$

(e) $\tau = 12.8$

(f) $\tau = 12.8$

(g) $\tau_{max} = 259.7$

(h) $\tau_{max} = 227.1$

(i) $\tau_{max} = 64.3$

Fig. 6.5: Iso-surface of $\alpha = 0.5$ showing effect of the Froude number on the temporal evolution of the cavity and lamella rim during drop impact on a liquid layer: $We = 1787$, $Oh = 0.0161$, $D = 0.175\ mm$, $V = 15\ m/s$

of Froude number augments significantly at the time instance when the maximum cavity depth is approached. As depicted in Figs 6.5g-6.5i, the cavity shape evolves from a V-shape towards a more hemispherical U-shape with decreasing Froude numbers i.e. increasing gravitational forces. A decrease of the Froude number leads to a considerably smaller maximum cavity depth and a sooner retraction of the cavity. For low Froude numbers, the velocities of the cavity expansion and retraction have the same order of magnitude. This can be seen by the almost symmetrical shapes of the cavity depth evolution curves in Fig. 6.6a for the two lower Froude numbers. The comparison with the analytical model of Bisighini et al. (2010) for the temporal evolution of the cavity depth shows rather good agreement for the cases with low Froude number whereas significant larger differences in the cavity depth evolution for large Froude number are found. Their model is derived from millimeter-sized drop impingement cases, hence, from impingement conditions with relatively low Froude numbers compared to the impacts of sprays relevant in practical situations. This may explain the discrepancies with their analytical model. As indicated by the graphs in Fig. 6.6a, it over-predicts the maximum cavity depth and assumes similar expansion and retraction velocities for the whole range of Froude numbers. This is however in contrast to the results of the numerical simulations. A reduction of the Froude number from $Fr = 131061$ to $Fr = 2571$, for instance, leads to a considerable decrease of the maximum cavity depth from $\Delta = 6$ to $\Delta = 3.6$, and moreover to smaller cavity retraction velocities. The reason for the smaller cavity depths and the reduced shape factor is related to the increasing role of gravitational forces inside the liquid film compared to the inertia (see Fig. 6.6b).

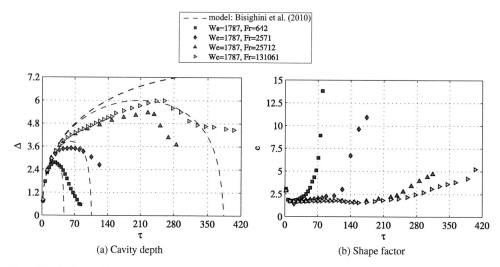

(a) Cavity depth   (b) Shape factor

Fig. 6.6: Effect of Froude number on the evolution of the cavity during single drop impingement onto a deep liquid pool: $We = 1787$, $Oh = 0.0161$, $D = 0.175\ mm$, $V = 15\ m/s$

In this paragraph, a thorough analysis of the isolated influence of the Froude number on the cavity evolution has been carried out. The theoretical model of Bisighini et al. (2010) proved to be accurate only within the parameter range for which it had been derived: it is not suitable for representing the temporal evolution of the cavity depth and the maximum cavity depth for engine-relevant Froude numbers. There is hence a need for an accurate modelling of the cavity penetration for the impingement conditions close to those appearing in bearing chambers. Due to the relevance of the centrifugal and gravitational forces acting on the droplets and wall films in bearing chambers, the role of the Froude number on the cavity penetration as detected in this subsection is significant. The results showed that the impact of a droplet onto the wall films in bearing chambers with large centrifugal acceleration and low gravitational forces (e.g. thin films near the Vent Offtake) may damp the cavity penetration and hence also the generation of secondary drops. It needs therefore to be taken into account, for the prediction of the influence of the height of the liquid film in the drop-film interaction modelling.

6.1.3   Effect of the Impinging Weber number on the Cavity Penetration

In this paragraph, the influence of the impinging drop velocity on the cavity evolution during single drop impingements will be analysed. The role of this influence will be determined while selecting a typical engine-relevant impingement scenario, where only the terminal velocity will be varied between $7\ m/s$ and $15\ m/s$. All the remaining parameters will be kept constant and summarized in Table 5.1 in Chapter 5.

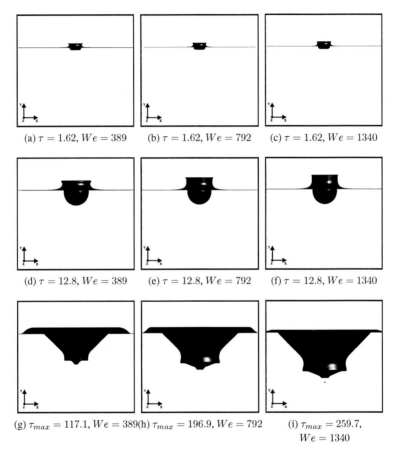

(a) $\tau = 1.62, We = 389$ (b) $\tau = 1.62, We = 792$ (c) $\tau = 1.62, We = 1340$

(d) $\tau = 12.8, We = 389$ (e) $\tau = 12.8, We = 792$ (f) $\tau = 12.8, We = 1340$

(g) $\tau_{max} = 117.1, We = 389$(h) $\tau_{max} = 196.9, We = 792$ (i) $\tau_{max} = 259.7,$
$We = 1340$

Fig. 6.7: Iso-surface of $\alpha = 0.5$ showing the effect of impact velocity on the temporal evolution of the cavity: $Oh = 0.0161, D = 0.175\ mm$

However, a change of the terminal impingement velocity implies not only a change in the Weber number, but also a change in the Froude number. Thus, increasing the terminal impingement velocity leads to an increasing dominance of the inertial forces over the surface tension, and the gravitational/centrifugal forces at the same time. Both overlapping effects will significantly be influence the cavity evolution. However, as investigated in the last section, the effect of the Froude number becomes noticeable only if a significant change of several orders of magnitude occurs. In this section, the impinging Weber numbers are varied in a range in which they are considered to have a much greater impact on the cavity evolution than does a change in the Froude number. Therefore, the results suit well the characterization of the influence of the impinging Weber number on the cavity evolution.

The results are illustrated at different time instances in Fig. 6.7 and accompanied by the quantitative temporal evolution of the cavity and lamella rim in Fig. 6.8.

Figs 6.7a-6.7c and 6.7d-6.7f present the effect of the Weber and Froude numbers on the initial stage of the impingement, namely the cavity formation and expansion regime. The evolution of the crater depth for different impingement Weber numbers in Fig. 6.8a shows equivalent crater expansion velocities in this regime for all the investigated Weber numbers. In the initial stage, the evolution of the shape factor in Fig. 6.8b appears unaffected by the Weber number.

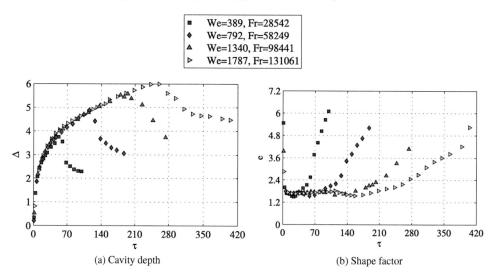

(a) Cavity depth   (b) Shape factor

Fig. 6.8: Effect of Weber number on the evolution of the cavity: $Oh = 0.0161$, $D = 0.175\ mm$

The expansion of the cavity in the general description in Section 6.1.1, continues until the conversion of kinetic energy into surface tension, potential energy, and dissipated energy, is completed, and the maximum cavity depth is reached (see Figs 6.7g-6.7i for the cavity depth). A typical maximum cavity depth, for instance, results for the case with the largest Weber number with $\Delta \approx 6$. From here on, a retraction of the cavity is initiated. The shape of the cavity depth evolution curves appears similar for all the investigated Weber numbers: simply scaled.

A reduction of the Weber number leads to a sooner retraction of the cavity. This behaviour is well depicted in Fig. 6.8a and qualitatively in Figs 6.7g-6.7i. The shape factor in Fig. 6.8b shows also an influence of the Weber number but only in the later impact evolution.

In conclusion, the Weber number has a significant effect on the cavity evolution and needs to be taken into account in the modelling of wall-effects later in this thesis. A rise in the Weber number equals an increase in the dominance of inertia (droplet and liquid film) over the surface tension forces. With respect to the cavity dynamics, the Weber number is directly proportional to the cavity depth and the cavity diameter. Hence, it is evident that larger impinging Weber numbers will result also in an enhanced cavity-wall interaction and a greater effect of the wall on the impingement evolution.

6.1.4   Effect of the Ohnesorge Number on the Cavity Penetration

In this section, the effect of the viscosity of the liquid on the cavity evolution is characterized for the parameter range for the aimed at applications to oil systems. It is examined using three different fluids. Here, the viscosity is varied between $\mu_l = 0.00102 \frac{kg}{ms}$ and $\mu_l = 0.015 \frac{kg}{ms}$. All remaining quantities are kept constant and listed in Table 5.3 in Chapter 5. In accordance with the results of the previous section, a qualitative description of the underlying effects is depicted in Fig. 6.9 and accompanied by the related temporal evolution curves in Fig. 6.10.

At the time instances when the cavity reaches its maximum extension, a noticeable influence of μ_l exists that is reflected by a reduction of the maximum cavity depth with increasing viscosity from $\Delta \approx 6$ to $\Delta \approx 4$ and differences in the cavity shape. Not only the cavity depth expansion velocity but also the cavity retraction phase seems to take place in a different manner for an increase in fluid viscosity. The cavity retraction starts much earlier for the case with the lowest fluid viscosity. It approaches a constant value in the further, but retracts later and with a larger retraction velocity for the cases with the two higher fluid viscosities. This difference in the retraction dynamics is also represented by the shape factor in Fig. 6.10b. Hence, with respect to the cavity depth retraction stage there is a transitional regime caused by the change in the ratio between the viscous forces and the surface tension forces. The existing pressure discontinuity along the inner surface of the cavity appears much less pronounced with an increase in viscosity. No capillary waves travelling along the surface can be identified for the larger liquid viscosities. However, the non-dimensional time to reach the maximum cavity depth and the initiation of the cavity retraction is within the same range for all investigated cases (see Figs 6.10a).

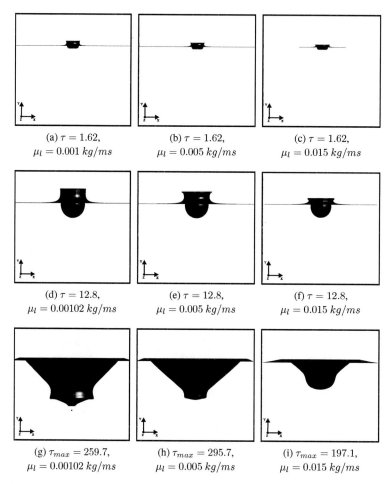

(a) $\tau = 1.62$,
$\mu_l = 0.001\ kg/ms$

(b) $\tau = 1.62$,
$\mu_l = 0.005\ kg/ms$

(c) $\tau = 1.62$,
$\mu_l = 0.015\ kg/ms$

(d) $\tau = 12.8$,
$\mu_l = 0.00102\ kg/ms$

(e) $\tau = 12.8$,
$\mu_l = 0.005\ kg/ms$

(f) $\tau = 12.8$,
$\mu_l = 0.015\ kg/ms$

(g) $\tau_{max} = 259.7$,
$\mu_l = 0.00102\ kg/ms$

(h) $\tau_{max} = 295.7$,
$\mu_l = 0.005\ kg/ms$

(i) $\tau_{max} = 197.1$,
$\mu_l = 0.015\ kg/ms$

Fig. 6.9: Iso-surface of $\alpha = 0.5$ showing the effect of liquid viscosity on the temporal evolution of the cavity and lamella rim during drop impact on a liquid layer: $We = 1787$, $Fr = 131061$, $D = 0.175\ mm$, $V = 15\ m/s$

In summary, it has been demonstrated for the kinematic impact conditions close to those of the oil systems of aero-engines that the effect of the viscosity of the liquid is decisive for the cavity evolution during a single drop impact onto a deep liquid pool. These findings have not been reported in the literature for large millimetre drops with capillary-driven impingement dynamics. An increase in viscosity results in smaller maximum cavity depths, not, however, affecting the duration of the impingement. The shape of the cavity changes by damping the pressure discontinuity and the capillary waves with higher viscosity. In bearing chambers, hence, an effect of the wall-film depth at the bounding walls on the secondary drop generation can significantly change from the conditions of cold start of the engine to warm operating conditions, even if the droplets have the same impinging kinetic energy and see the same wall-film thickness

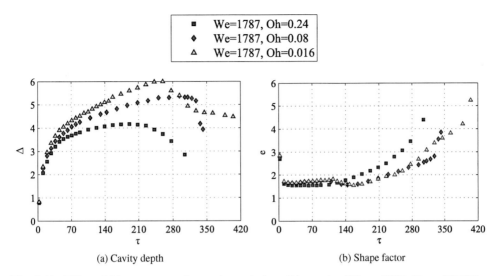

(a) Cavity depth   (b) Shape factor

Fig. 6.10: Effect of Ohnesorge number on the evolution of the cavity: $We = 1787$, $Fr = 131061$, $D = 0.175\ mm$, $V = 15\ m/s$

before impacting. The current drop-film interaction models are not capable of describing this effect. Thus, it must be considered in the later drop-film interaction modelling chapter.

6.1.5 Effect of Impingement Angle on the Cavity Penetration

The effect of the impingement angle will be studied analogously to the previous effects by using a typical engine-relevant kinematic impingement condition with $We = 541$, $Oh = 0.0088$, and $Fr = 131061$, while only varying the impingement angle, within the expected range, namely between $\alpha = 90°$ and $\alpha = 30°$. The test cases of the present study are summarized in Table 5.4 in Chapter 5. The iso-surface of the liquid volume fraction at $\alpha = 0.5$ in Fig. 6.11 at three different impingement times depicts the impingement evolution. Quantitative results of the cavity expansion are presented in the graphs in Fig. 6.12. The cavity formation regime, at $\tau = 1.62$, is shown in Figs. 6.11a- 6.11c. A reduced impingement angle has as a consequence that only a fraction of the cavity is being formed. The smaller the impingement angles the smaller the size of the cavity which is formed. Hence, this already indicates the strong dependence of the cavity depth evolution on the normal velocity of the impinging droplet. In addition, the cavity appears to evolve at an inclination compared to the normal impingement in the direction of the absolute velocity vector.

In contrast to the normal impingement, in oblique impingements, the cavity shape at $\tau = 31$ is not hemispherical. The level of sphericity is here considerably reduced with smaller impingement angles. The lowest point of the cavity is not located at the center of the cavity, as it is in normal impingement, but is shifted downstream from the impingement point. By building the non-dimensional time scale τ using only the normal impingement velocity in the quantitative analysis of the cavity depth and shape in the Figs 6.12a and 6.12b, an excellent fit of the different curves is obtained. This endorses the aforementioned strong dependence of the cavity depth evolution on the normal impinging velocity. At the time instance when the maximum cavity depth is reached, the differences in cavity shape become even more evident. Not only is the level of inclination increased, but also the maximum cavity depth is reduced considerably at flatter impingement angles. The time to reach the maximum cavity depth is also affected by the impingement angle. The graphs in Fig. 6.12a reveal, for instance, for an impingement angle of $\alpha = 30°$ a maximum cavity depth of $\Delta_{max} \cong 2.4$ at $\tau_{max} \cong 20$. For the normal impingement, the maximum cavity depth is reached with $\Delta_{max} \cong 4.2$ at $\tau_{max} \cong 80$.

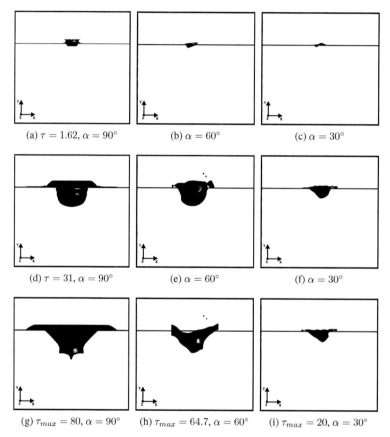

(a) $\tau = 1.62$, $\alpha = 90°$ (b) $\alpha = 60°$ (c) $\alpha = 30°$

(d) $\tau = 31$, $\alpha = 90°$ (e) $\alpha = 60°$ (f) $\alpha = 30°$

(g) $\tau_{max} = 80$, $\alpha = 90°$ (h) $\tau_{max} = 64.7$, $\alpha = 60°$ (i) $\tau_{max} = 20$, $\alpha = 30°$

Fig. 6.11: Iso-surface of $\alpha = 0.5$ showing the effect of impingement angle on the temporal evolution of the cavity and lamella rim during drop impact on a liquid layer: $We = 541$, $Fr = 131061$, $Oh = 0.0088$

The cavity expansion and retraction velocity is in very good agreement for the different impingement angles which makes the shape of the curves very similar to those found in Section 6.1.3 for the effect of the Weber number. The main differences in the cavity retraction regime between normal and oblique impingements are mainly manifested in terms of the retraction initiation.

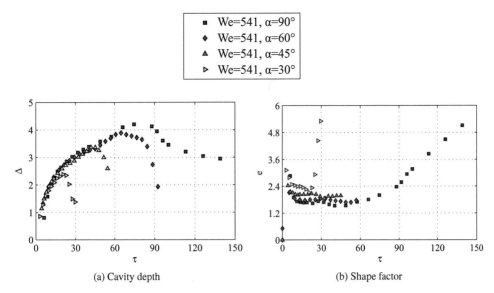

(a) Cavity depth   (b) Shape factor

Fig. 6.12: Effect of impingement angle on the evolution of the cavity: $We = 542$, $Fr = 131061$, $Oh = 0.0088$

By observing both, in Figs 6.12a and 6.12b, a trend is noticeable showing sooner retraction at smaller impingement angles.

From the results of the present section, it can be concluded that the effect of the impingement angle on the cavity expansion is considerable. The cavity depth expansion for oblique impingements is proportional to the normal impinging kinetic energy, hence, to the normal Weber number. This implies that the effect of the impingement angle on the maximum cavity depth is well described using normal impingements scenarios with corresponding (normal) Weber numbers. The shape and expansion direction of the cavity, however, differ significantly from the case of normal impingements. It expands at an inclination to the direction of the absolute impinging velocity and not with a hemispherical shape. The effect of the wall on the secondary drop generation for sprays impacting with equivalent kinetic energy hence decreases significantly with flatter impingement angles, since the cavity penetration reduces with the smaller normal velocity component. The effect of the wall is hence decisively influenced by the impingement angle of the sprays and needs therefore to be considered in the drop-film interaction modelling.

With the presented numerical results in hand, knowledge about the maximum cavity penetration subject to different effects during deep pool impact has been achieved. Hence, with respect to the characterization of the influence of the wall-film height on the secondary drop generation, all details are available. In the next section, an analysis and relation between the combined effect of the relative film height and the maximum cavity penetration will be analysed.

6.2 Effect of Cavity Penetration & Liquid Film Height on the Secondary Drop Generation

In this section, the aforementioned relation between the cavity penetration, the lamella expansion and the secondary drop generation for a quantitative characterization of the wall effects in drop impact dynamics is analysed and discussed. The section will show how the maximum cavity depth for deep pool impingements can be used as the governing parameter together with the relative film height H^* to quantify the wall effects.

6.2.1 Relation between Cavity Penetration and Lamella Expansion

In the present section, the relation between the cavity penetration and the lamella expansion is determined by using four drop impingement scenarios with different relative film heights of $H^* = 1$, $H^* = 3$, $H^* = 6$, $H^* >> 6$. As already investigated in Bisighini (2010), the ratio $H^*/\Delta_{max,deep}$ is determined as the crucial parameter representing the magnitude of the wall effects taking place in each drop impingement. This is at least the case for thin film and shallow pool impingement where the formation of a cavity inside the liquid wall film is clearly distinguishable. As shown in the previous sections, the maximum cavity depth resulting for a normal impingement onto a deep pool only depends on the Weber number, the Froude number and the Ohnesorge number. These quantities are constant for all the investigated cases in this section. Thus, the maximum cavity depth for a deep pool impingement is approximately $\Delta_{max,deep} \approx 6$ for all cases in this section. According to the previous sections, results of the numerical simulations are discussed for typical engine-relevant impingements. The kinematic impact conditions and the fluid properties are again identical and listed in Table 6.1.

case	We	Fr	Oh	$H^*/\Delta_{max,deep}$
1	1787	131061	0.016	0.167
2	1787	131061	0.016	0.5
3	1787	131061	0.016	1
4	1787	131061	0.016	∞

Table 6.1: Test conditions for the effect of $H^*/\Delta_{max,deep}$ on the cavity and lamella expansion

The experimental results of Bisighini (2010) already showed for millimetre sized drop impingements that the liquid film influence becomes visible when the cavity reaches and senses the wall. Figs 6.13a-6.13d confirm their findings since no effect of the liquid film height on the cavity formation and lamella rim expansion is indicated in the very early stage of the impingement at $\tau = 1.62$.

In the further evolution, when the cavity of the case with the smallest value of $H^*/\Delta_{max,deep} = 0.167$ in Fig. 6.13e at $\tau = 12.8$ senses the wall, an effect on the lamella height first becomes visible. At this instance, the cavity can no more extend in depth. The impinging kinetic energy not needed for the further cavity penetration is redirected and used to further expand the lamella.

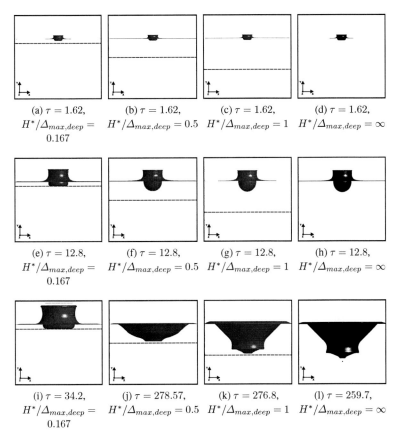

(a) $\tau = 1.62$,
$H^*/\Delta_{max,deep} = 0.167$

(b) $\tau = 1.62$,
$H^*/\Delta_{max,deep} = 0.5$

(c) $\tau = 1.62$,
$H^*/\Delta_{max,deep} = 1$

(d) $\tau = 1.62$,
$H^*/\Delta_{max,deep} = \infty$

(e) $\tau = 12.8$,
$H^*/\Delta_{max,deep} = 0.167$

(f) $\tau = 12.8$,
$H^*/\Delta_{max,deep} = 0.5$

(g) $\tau = 12.8$,
$H^*/\Delta_{max,deep} = 1$

(h) $\tau = 12.8$,
$H^*/\Delta_{max,deep} = \infty$

(i) $\tau = 34.2$,
$H^*/\Delta_{max,deep} = 0.167$

(j) $\tau = 278.57$,
$H^*/\Delta_{max,deep} = 0.5$

(k) $\tau = 276.8$,
$H^*/\Delta_{max,deep} = 1$

(l) $\tau = 259.7$,
$H^*/\Delta_{max,deep} = \infty$

Fig. 6.13: Iso-surface of $\alpha = 0.5$ showing the effect of liquid film height on the temporal evolution of the cavity and lamella rim during drop impact on a liquid layer: $We = 1787$, $Fr = 131061$, $Oh = 0.0161$: $D = 0.175\ mm$, $V = 15\ m/s$

This leads to a larger lamella rim height and diameter. This becomes clearly visible in Fig. 6.14 after $\tau \approx 12$. The maximum lamella rim height, for instance, results with $H_{cr}/D \approx 3.4$ at $\tau_{max} \approx 36$ compared to $H_{cr}/D \approx 2.5$ at $\tau_{max} \approx 28$ for the deep pool impact. For the impingement onto a liquid film with $H^*/\Delta_{max,deep} = 0.167$, the strong displacement of the liquid film causes moreover a complete dry out. Here, the maximum cavity depth equals the initial film height ($H^* = \Delta_{max} = 1$) while further expanding in horizontal direction. The further horizontal expansion is visible in Fig. 6.14b which shows a further increase of the shape factor in time. Similar dynamics take place some time later when the cavity of the case $H^*/\Delta_{max,deep} = 0.5$ reaches the bottom wall. Accordingly, a plateau at the center of the cavity evolves with ($H^* = \Delta_{max} = 3$) which is illustrated in Fig. 6.14a. In this case, however, a much larger portion of kinetic energy has already been transferred to surface tension and dissipation compared to the thin film impact. Because of the smaller amount of kinetic energy redirected from the wall a considerably less pronounced increase in the shape factor and the lamella rim

height is observed. On the other hand the time to reach the maximum lamella rim height is
slightly increased compared to the case with $H^*/\Delta_{max,deep} = 0.167$. The lamella rim diameter
appears in this case ($H^*/\Delta_{max,deep} = 0.5$) unaffected by the change of the liquid film height. The
pressure discontinuity caused by the falling back of the lamella rim is strong enough to overcome
the viscous forces and form capillary waves that propagate to the centre of the cavity and change
its shape from a hemisphere to a cone (see Fig. 6.13f).

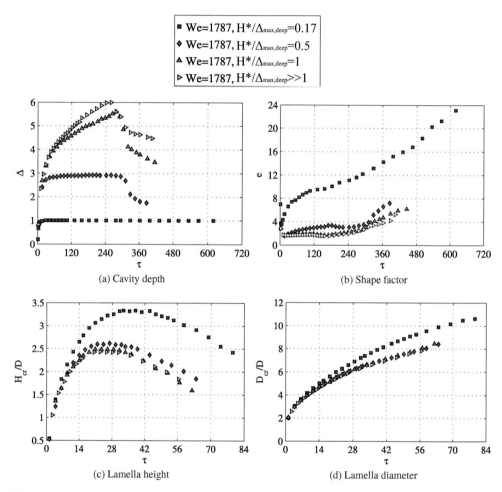

Fig. 6.14: Effect of liquid film height on the evolution of the cavity and lamella rim during single
drop impingement onto a deep liquid pool: $We = 1787$, $Fr = 131061$, $Oh = 0.0161$:
$D = 0.175\ mm$, $V = 15\ m/s$

A further increase in liquid film height to $H^*/\Delta_{max,deep} = 1$, corresponding to the maximum
cavity depth of the deep pool impact, causes a reduction of the cavity depth expansion velocity.
The maximum cavity depth is slightly reduced compared to the deep pool impingement and
reaches the wall after the propagation of the capillary waves is initiated and the shape is changed to

a conical shape. The shape factor, the lamella rim height and the lamella rim diameter evolution are nearly identical with the deep pool impingement. The cavity and lamella rim retraction velocity is also similar for all investigated test cases. For the cavity depth evolution, additionally, the cavity retraction is also initiated at the same time instances.

In conclusion, the wall (liquid film height) has a considerable impact on the evolution of the cavity and the lamella rim. The results in this range of Froude and Capillary numbers are in good agreement with the findings of van Hinsberg (2010) and Bisighini (2010) for large millimetre sized drop impingements. The impingement of single drops onto liquid surfaces sense the wall only if the maximum cavity depth reaches the bottom wall. This confirms the fact that the influence of the liquid film height is not sufficiently characterized by only the relative film height H^* as it has been proposed in nearly all previous spray-film interaction models.

An improved classification of the wall influence during single drop impingement onto thin, shallow and deep liquid pool will be conducted next. For drop impingement onto liquid film heights having the ratio $H^*/\Delta_{max,deep} < 0.5$, the effect of the wall on the lamella rim evolution is significant. This impingement may hence be referred to as thin film impingement. For $H^*/\Delta_{max,deep} > 0.5$, on the other hand, this effect is not significant as confirmed by the impingement onto liquid surfaces with $H^* > 3$ and thereby be classified as a shallow pool impingement. In the next section, this ratio will be used to quantitatively describe the effect of the wall on the secondary drop generation. It requires the knowledge of the maximum cavity depth for deep pool impingement that has been obtained in this sub-section.

6.2.2   Effect of the Ratio $H^*/\Delta_{max,deep}$ on Secondary Drop Generation

With the knowledge gained from the previous sections describing the cavity penetration during drop impingement, the ratio $H^*/\Delta_{max,deep}$ can now be used for the evaluation of the liquid film influence on the secondary drop generation. The influence of $H^*/\Delta_{max,deep}$ on the secondary drop characteristics and ejected mass fraction will be studied for two different reference test cases, namely for $We = 350$ and $We = 1850$. These ranges of Weber numbers represent both large millimeter drops (cases 4-7) and a technical relevant submillimeter drop impingement (cases 1-3). By obtaining the values of $\Delta_{max,deep}$ from the cavity expansion dynamics in the previous sections ($\Delta_{max,deep} = 2.53$ for $We = 350$ and $\Delta_{max,deep} = 5.83$ for $We = 1850$), different effects of the bottom wall will be examined. The test cases are listed in Table 6.2.

case	We	Fr	Oh	α	$H^*/\Delta_{max,deep}$
1	1850	151833	0.036	90	0.17
2	1850	151833	0.036	90	0.51
3	1850	151833	0.036	90	1
4	350	295	0.002	90	0.04
5	350	295	0.002	90	0.4
6	350	295	0.002	90	0.8
7	350	295	0.002	90	1

Table 6.2: Test conditions for the effect of liquid film height

Fig. 6.15 shows the influence of $H^*/\Delta_{max,deep}$ on the crown's breakup dynamics for an impingement of $We = 1850$ onto liquid surfaces. The influence is depicted for three different stages during the impingement evolution. Accordingly, the quantitative analysis of the secondary drop characteristics and ejected mass fraction will be displayed in Fig. 6.16. At the initial stage of the impingement ($\tau = 2.3$), no effect of the reduced film thickness on the crown's dynamics is visible in Figs 6.15a-6.15c. This may be explained by the fact that the cavity is still in its expansion phase and does not sense the bottom wall. During the further impingement evolution shown in Figs 6.15d-6.15f, an influence of the wall is present only for $H^*/\Delta_{max,deep} = 0.17$. Here, the remaining kinetic energy, that cannot be invested into a further expansion of the cavity, is redirected and transferred to the crown leading at this stage to a slightly larger crown height compared to the cases with larger film heights. The crown shape is however in good agreement with larger film thickness impacts. The bottom wall effect is more pronounced at the later stages of the impingement, at $\tau = 8.8$, where a significant increase of the crown height occurs for $H^*/\Delta_{max,deep} = 0.17$ (see Figs 6.15g-6.15i). The two larger film heights are still independent of the film thickness since neither cavity bottom contacts the wall. When observing the secondary drop characteristics, it is also seen that for the shallow film case, larger secondary drops are released from the crown's rim, and there is an increased crown height as well.

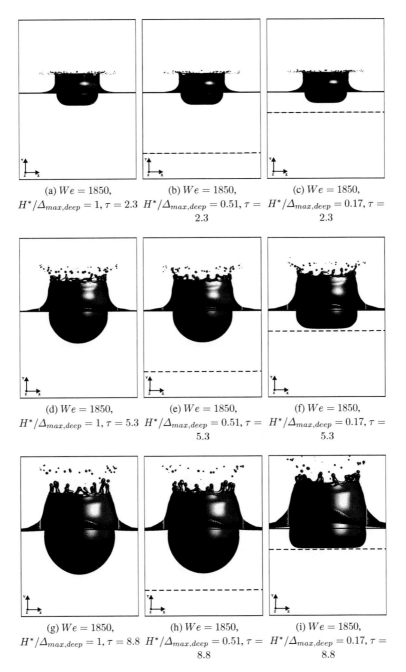

(a) $We = 1850$, $H^*/\Delta_{max,deep} = 1, \tau = 2.3$

(b) $We = 1850$, $H^*/\Delta_{max,deep} = 0.51, \tau = 2.3$

(c) $We = 1850$, $H^*/\Delta_{max,deep} = 0.17, \tau = 2.3$

(d) $We = 1850$, $H^*/\Delta_{max,deep} = 1, \tau = 5.3$

(e) $We = 1850$, $H^*/\Delta_{max,deep} = 0.51, \tau = 5.3$

(f) $We = 1850$, $H^*/\Delta_{max,deep} = 0.17, \tau = 5.3$

(g) $We = 1850$, $H^*/\Delta_{max,deep} = 1, \tau = 8.8$

(h) $We = 1850$, $H^*/\Delta_{max,deep} = 0.51, \tau = 8.8$

(i) $We = 1850$, $H^*/\Delta_{max,deep} = 0.17, \tau = 8.8$

Fig. 6.15: Qualitative depiction of the liquid film height influence on the crown's breakup process during single drop impingement onto liquid surfaces

The quantitative analysis in Fig. 6.16 supports the previous observations since it shows for both Weber numbers a visible increase in the mass and number of the secondary drops for values of $H^*/\Delta_{max,deep} < 0.4$ (see Figs 6.16a and 6.16b). The transition from a film thickness of $H^*/\Delta_{max,deep} = 0.51$ to $H^*/\Delta_{max,deep} = 0.17$ causes hence an enhancement of the secondary drop mass, from $M_{sec}/M \approx 0.175$ to $M_{sec}/M \approx 0.225$, and an increase of the droplet number, from $N \approx 150$ to $N \approx 205$. For $0.4 < H^*/\Delta_{max,deep} < 1$, no effect of the bottom wall is detected. This enforces the fact that significant damping exists at larger film heights due to the investment of more impinging kinetic energy into the cavity expansion. Bisighini (2010) found qualitatively similar thresholds between a thin film and a thick film impingement in his investigation. The present outcome is in contrast to the results of Okawa et al. (2006). They found a dependence on H^* only for the number of secondary drops but not for the ejected mass fraction. Their correlations were developed by using the K-number only although having a large variation of the impinging Froude numbers. For this reason and in accordance with the arguments made in the present study, a cross-effect of the role of gravity may exist that makes their correlations questionable. On the other hand, they used only the relative film thickness H^* for quantification of the wall influence, something which was proved here to be insufficient.

The arithmetic mean secondary drop sizes in Fig. 6.16c demonstrate furthermore that the additional release of secondary drop mass caused by the bottom wall effect is mostly due to large secondary drops released at later stages of the impingement. This is supported by an increase of the arithmetic mean secondary drop size and logarithmic standard deviation for $H^*/\Delta_{max,deep} < 0.4$. For larger relative film heights up to $H^*/\Delta_{max,deep} = 1$, secondary drop size and logarithmic standard size distribution are independent of $H^*/\Delta_{max,deep}$.

The additional acceleration of the crown in vertical direction due to the wall effects appears to increase also the orthogonal mean secondary drop velocities (see Fig. 6.16f for $H^*/\Delta_{max,deep} < 0.4$). The parallel component of the mean absolute velocity is independent of the $H^*/\Delta_{max,deep}$ parameter. Hence, there are also steeper ejection angles of the secondary drops expected as soon as the impingement event senses the wall.

The present section clearly supports the statement of Bisighini (2010) and the study of the cavity/lamella rim expansion presented in Section 6.1, asserting that a thorough analysis of the bottom wall effect is only possible with the parameter $H^*/\Delta_{max,deep}$. All other studies using only the relative film thickness H^* for analysing the wall effects are overestimating the effect of liquid film height particularly for lower impingement energy. A quantitative analysis of the secondary drop characteristics and ejected mass fraction using the parameter $H^*/\Delta_{max,deep}$ was not addressed in literature. The ratio has been shown here to have a significant influence on the secondary drop characteristics and ejected mass fraction for values of $H^*/\Delta_{max,deep} < 0.4$, since increase of mass, number, arithmetic mean diameter, logarithmic standard deviation and orthogonal velocity of secondary drops were resolved by the 3D-VOF-AMR method. For larger values of $H^*/\Delta_{max,deep}$, up to the maximum value of 1, the effect of the wall on the secondary drop characteristics and ejected mass fraction is mostly insignificant. Note that the identified trend is only valid down to a minimum value of $H^*/\Delta_{max,deep} \approx 0.04$. For values smaller than 0.04, the effect of the bottom wall may differ from the trends identified here since the effects of surface roughness and wettability may change the dynamics.

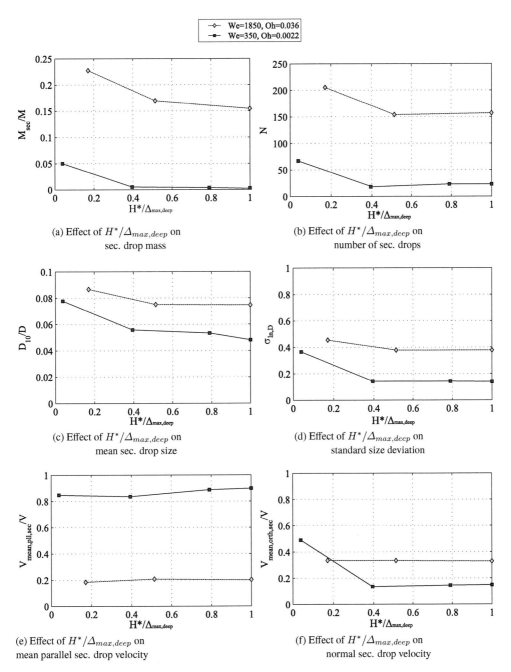

Fig. 6.16: Effect of liquid film height on characteristics and mass of secondary drops during single drop impact onto liquid surfaces

With the numerical results gained in this and the previous chapter, a number of the most dominant parameters influencing the secondary drop generation during single drop impingements were investigated. The numerical and experimental database developed thereby fulfill all the requirements for the derivation of suitable correlations to form a drop-film interaction model applicable into simulation tools for oil system flows in aero-engines.

7 Correlations for the Products of Splashing

With the help of the developed data base in the experimental validation chapter 4, and the numerical simulations of drop impingement with and without wall effects in chapter 5 and 6, a complete data base has been developed that enables the derivation of several correlations. The correlations will form the drop-film interaction model that can be integrated in the simulation methods describing the oil flow system in aero-engines. However, before leading to the correlation development, a brief description of the pre-impingement parameters and the target quantities for a drop-film interaction model is given.

7.1 Requirements for the Drop-Film Interaction Model

Based on the numerical results gained in the previous chapters and the review of scientific literature, several requirements for an improved drop-film interaction model can be defined. A drop-film interaction model for application in oil flow systems needs to evaluate the mass and momentum balance between the impinging drops and the liquid wall film.

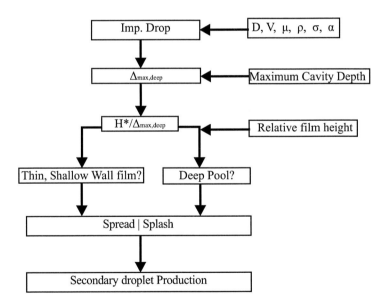

Fig. 7.1: Definition of pre -and post impingement parameters needed in a drop-film interaction model

The mass exchange between impinging droplets and liquid wall-film can be determined by predicting the mass of the ejected secondary spray and relating it with the impacting droplet mass, M_{sec}/M. On the other hand, the momentum exchange requires the knowledge of the momentum of the impinging and the ejected secondary droplets (primary and secondary drop velocities).

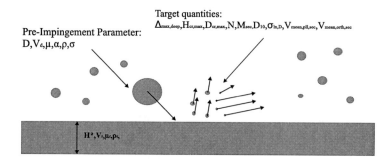

Target quantities:
$\Delta_{max,deep}, H_{cr,max}, D_{cr,max}, N, M_{sec}, D_{10}, \sigma_{ln,D}, V_{mean,pll,sec}, V_{mean,orth,sec}$

Pre-Impingement Parameter:
$D, V_d, \mu, \alpha, \rho, \sigma$

$H^*, V_f, \mu_f, \rho_f,$

Fig. 7.2: Sketch showing the pre -and post impingement parameter

Hence, it needs to predict the target secondary drop quantities depending on the non-dimensional pre-impingement drop parameter. In the Figs 7.1 and 7.2, the main pre -and post impingement parameters needed for the formulation of a drop-film interaction model are listed.

Before a droplet impinges on a wall film, there are several impinging parameters that need to be determined, i.e. the impinging drop diameter, velocity, fluid properties, impingement angle, as well as the centrifugal and the gravitational acceleration. Based on this information, the maximum cavity depth for an equivalent deep pool impact has to be determined. The maximum cavity depth combined with the relative film height will then classify the impingement into a thin, shallow or deep pool impingement. Once all relevant impinging parameters are collected, the next step is the evaluation of the secondary drop characteristics and ejected mass fraction. Here, first the release position of the secondary drops is defined using the maximum lamella height and diameter at a certain non-dimensional time. In this position, then, the secondary droplet size and velocity distribution, number and mass is evaluated. By adopting this process for each impinging droplet in a flow field and essentially for sparse spray impact characteristics, prediction of the secondary spray can be determined. The correlations to be derived in this section and in the subsequent sections will be based on a superposition of all the following pre-impingement quantities of influence

- the normal Weber number We_n,
- the Weber number We,
- the Ohnesorge number Oh,
- the Froude number Fr,
- the parameter $H^*/\Delta_{max,deep}$,
- the impact angle α.

The target quantities are correlated all in the same manner using the superposition principle of each influence parameter. This is accomplished by assigning each parameter of influence an exponent where the value states its dominance in the impingement dynamics. Two general equations are defined, one for the cavity penetration and lamella expansion, P_{CL}, and one for the secondary drop characteristics and ejected mass fraction, P_{SD}. The equation P_{CL} accounts for the impingement angle by using only the normal Weber number, We_n. The equations are defined in the exemplary form

$$P_{CL} = a \cdot We_n^b \cdot Oh^c \cdot Fr^d \cdot (\frac{H^*}{\Delta_{max,deep}})^e + f, \tag{7.1}$$

where P_{CL} is the target quantity describing the maximum cavity and lamella evolution, a is a coefficient, b,c,d,e are exponents, f a constant, and

$$P_{SD} = a \cdot We^b \cdot Oh^c \cdot Fr^d \cdot (\frac{H^*}{\Delta_{max,deep}})^e \cdot \alpha^f + g. \tag{7.2}$$

Here, P_{SD} is the target quantity describing the secondary droplet characteristics and ejected mass fraction, a is a coefficient, b,c,d,e,f are exponents and g a constant. All the coefficients, exponents and constants are obtained by applying a Least-Square fit between the correlation and the actual data. In order to make the correlation also applicable outside of the range covered by the experiments and the VOF-AMR predictions only linear and exponential functions were used. In the following section, all relevant parameters that were mentioned in this section will be correlated based on the developed data base in the previous chapters.

7.2 Correlations for the Maximum Cavity Penetration and the Lamella Rim Height

In chapter 6, wall effects in single drop impingement onto liquid surfaces have been described by using the ratio between the maximum cavity depth (for an equivalent deep pool impingement) $\Delta_{max,deep}$ and the relative film height H^*. Here, it was shown that wall-effects occur only if a cavity-wall interaction is predicted with $\Delta_{max,deep} < 0.5$. For the modelling of drop-film interaction, this section aims hence first to find suitable correlations for the prediction of the maximum cavity depth $\Delta_{max,deep}$ at $\tau_{cav,max,deep}$ during single drop impingement onto a deep liquid pool. With this in hand, the level of influence of the bottom wall on the drop impingement can be determined using the quantity $H^*/\Delta_{max,deep}$.

Maximum Cavity Depth

A correlation relating the non-dimensional impinging drop parameters to the maximum cavity depth is discussed in this section. The maximum cavity depth for deep pool impingements is directly proportional to the impinging drop kinetic energy.

In Fig. 7.3, the axisymmetric VOF-AMR results of the maximum cavity depth and the non-dimensional time to reach $\Delta_{max,deep}$ are compared to the predictions of the analytical model of Bisighini et al. (2010) and equation 7.1, respectively. The dashed line in both figures represents an agreement between the numerical simulation and the theoretical or correlated values. Here, the diamonds compare all the theoretical predictions of Bisighini et al. (2010) with the numerical

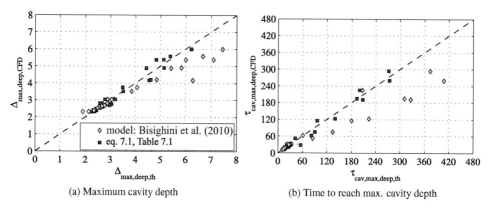

(a) Maximum cavity depth   (b) Time to reach max. cavity depth

Fig. 7.3: Maximum crater depth $\Delta_{max,deep}$ at $\tau_{cav,max,deep}$ in deep pool impact: Comparison of all simulation results with the theoretical model of Bisighini et al. (2010) and equation 7.1

P	a	b	c	d	e	f
$\Delta_{max,deep}$	0.00225	0.4985	-0.18466	0.2585	0	1.969
U_Δ	0	±0.036	±0.055	±0.018	0	0
$\tau_{cav,max,deep}$	0.000643	1.127	-0.0891	0.353	0	13.423
U_τ	0	±0.035	±0.088	±0.035	0	0

Table 7.1: Coefficients and exponents for the maximum cavity depth $\Delta_{max,deep}$ and $\tau_{cav,max,deep}$ including the uncertainty U_x in the powers

simulations. Accordingly, the squares depict a comparison between the correlation (eq. 7.1 and table 7.1) and the numerical simulations. The model of Bisighini et al. (2010) is selected here since it proved to represent the available experimental results of millimetre sized drop impingements onto deep liquid pool sufficiently well (see i.e. Engel (1966), Macklin and Metaxas (1976), Liow (2001), Brutin (2003) etc.). However, the predictions of the analytical model of Bisighini et al. (2010) evidence a significant over-prediction of the maximum cavity depth. This occurs in particular for large $\Delta_{max,deep}$ for drop impingements with elevated Froude numbers. This behaviour is confirmed for both, the maximum cavity depth and the time to reach the maximum cavity depth. On the other hand, the low Froude number impingements deliver a rather good comparison of the model of Bisighini et al. (2010) with the numerical simulations. This indicates again that the role of the external field forces is not captured correctly in his analytical model. The comparison of the developed correlations derived from equation 7.1 using the values from the table 7.1 show a considerable improvement. The latter is confirmed since the squares are well correlated with the dashed line in Figs 7.3 although at higher Froude numbers there is a weak scattering around the dashed line. The related uncertainty U_x in the powers of the correlations here and hereafter are also listed in the different tables (see, e.g. Table 7.1).

Maximum Lamella Rim Height and Diameter

An accurate prediction of the lamella rim expansion, namely the lamella rim height and diameter, is needed to model the position of secondary droplet release during each impingement event. The release position of secondary drops is needed to ensure accurate starting positions of the secondary droplets.

P	a	b	c	d	e	f
$H_{cr,max}/D$	0.000372	1	-0.19	0.05	-0.1	0
U_H	0	±0.015	±0.03	±0.01	±0.02	0
$\tau_{cr,max}$	0.00712	1	-0.1	0.04	-0.13	-2.9406
U_τ	0	±0.01	±0.025	±0.01	±0.02	0

Table 7.2: Coefficients and exponents for the maximum lamella rim height $H_{cr,max}/D$ and $\tau_{cr,max}$ including the uncertainty U_x in the powers

With the help of the superposition equation proposed in equation 7.1, two further correlations will be derived from the existing data base (including the cases with different liquid film depth) for the maximum lamella rim height $H_{cr,max}/D$ at $\tau_{cr,max}$. From these quantities the appropriate constants are listed in table 7.2. In Fig. 7.4, the comparison of the experimental and numerical results to the correlations for $H_{cr,max}/D$ and $\tau_{cr,max}$ is shown. A very good agreement is confirmed. The dashed line represents the theoretical predictions of equation 7.1 with constants from table 7.2. The proposed correlations are valid for normal drop impingement with $\alpha = 90°$. For non-normal drop impingements, it has been shown that a very complex and non-linear behaviour of the lamella height and diameter evolution exists. However, as a first approach, it is suggested to calculate the maximum lamella height of oblique impingements also using the outcome of normal impingements with equivalent (normal) Weber numbers. Based on the actual observation this approach may result in inaccurate results for steep impingement angles between $60° < \alpha < 90°$ but still in an acceptable range for all other impingement angles.

The prediction of the lamella rim diameter is determined using the model proposed by Yarin and Weiss (1995) and successfully adopted by Cossali et al. (2004). The coefficient C and the exponent n in their model, $D_{cr}/D = C \cdot (\tau - \tau_0)^n$, to fit the experimental and numerical results are shown in equation 7.3. The coefficient C is constant in all the fittings with 0.95. Interestingly, the influence of the wall on the lamella rim diameter evolution is also modelled by including the ratio $H^*/\Delta_{max,deep}$ within the exponent n (see equation 7.3).

In Fig. 7.4c, a rather good agreement between equation 7.3, the experimental and the numerical results is found. This is confirmed for thin film impingements, $H^*/\Delta_{max,deep} = 0.17$, and thick/deep pool impingements, $H^*/\Delta_{max,deep} = 1$, respectively.

$$D_{cr}/D = 0.95 \cdot (\tau - \tau_0)^{(0.34835 \cdot (\frac{H^x}{\Delta_{max,deep}} - 0.13543)^{-0.032585})}, \qquad (7.3)$$

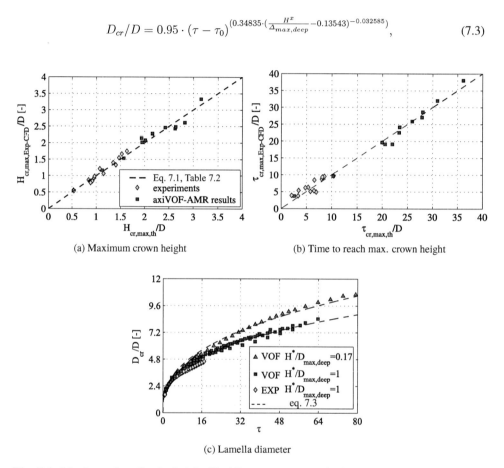

(a) Maximum crown height

(b) Time to reach max. crown height

(c) Lamella diameter

Fig. 7.4: Maximum lamella rim height H_{cr}/D at $\tau_{cr,max}$ in single drop impingement onto liquid surfaces $H^* > 1$: Comparison of all simulation and experimental results with the predictions of equation 7.1 and 7.3

7.3 Correlating the Secondary Drop Characteristics and Ejected Mass Fraction

According to the cavity and lamella expansion during single drop impact, different effects on the secondary drop characteristics and ejected mass fraction were thoroughly analysed and characterized, too. Hence, with respect to the drop-film interaction model to be developed in this thesis, it is necessary to summarize the results of the secondary drop characteristics and ejected mass fraction in a set of correlations depending only on quantities characterizing the pre-impingement. Hence, the present section aims at the development of suitable correlations describing the secondary drop characteristics and ejected mass fraction during single drop impingement.

Ejected Mass Fraction and Number of Secondary Drops

The most important parameter in spray-film interaction models is the determination of the ejected mass fraction of the impinging drop for different impinging drop parameter. By using all the experimental and the numerical results produced in this thesis, a trend for the ejected mass fraction M_{sec}/M could be found. Since the effect of the impingement angle showed to affect the splashing products with a non-linear behaviour, two set of correlation are developed, one for $45° < \alpha < 90°$ and one for $30° < \alpha < 45°$, respectively.

P	a	b	c	d	e	f	g
M_{sec}/M	0.0021092	1	-0.25	0.063	-0.04	-0.95	0.067961
U_M	0	±0.02	±0.05	±0.01	±0.01	±0.05	0
N	0.018703	1	-0.25	0.063	-0.22	0	-27.897
U_N	0	±0.02	±0.05	±0.01	±0.01	0	0

Table 7.3: Coefficients and exponents for the ejected mass M_{sec}/M and number N of secondary drops during single drop impingement onto liquid surfaces for $45° < \alpha < 90°$ including the uncertainty U_x in the powers

P	a	b	c	d	e	f	g
M_{sec}/M	0.0021092	1	-0.25	0.063	-0.04	-0.95,with $90 - \alpha$	0.067961
U_M	0	±0.02	±0.05	±0.01	±0.01	±0.05	0
N	$1.208 \cdot 10^{-5}$	1	-0.25	0.063	-0.22	2	-51
U_N	0	±0.02	±0.05	±0.01	±0.01	±0.01	0

Table 7.4: Coefficients and exponents for the ejected mass M_{sec}/M and number N of secondary drops during single drop impingement onto liquid surfaces $30° < \alpha < 45°$ including the uncertainty U_x in the powers

For impingements $\alpha < 30°$ no splashing regime is expected for the parameter range found in oil systems of aero-engines. Tables 7.3 and 7.4 list the coefficient, exponents and constant needed in equation 7.2 for a Least-Square fit of the experimental and numerical results. Figure 7.5a shows the validity of the correlation against the experimental and numerical results. Here, a rather good comparison is found. Appreciating the range of validity of these correlations reaching from millimetre sized drops to technical relevant submillimeter drop impingements, and the fact that measurement techniques to measure the mass of two-phase are limited; the weak scattering of the data around the correlation is acceptable. Accordingly, the number of secondary drops could also be correlated by equation 7.2, table 7.3 and 7.4 for both levels of impingement angles. The validity of this correlation is depicted in Figure 7.5b. This target parameter is considered to be very important since it offers the possibility to directly retrieve details about the volumetric mean diameter D_{30} of the produced secondary spray in advance which may be very useful in many occasions of spray impingement modelling.

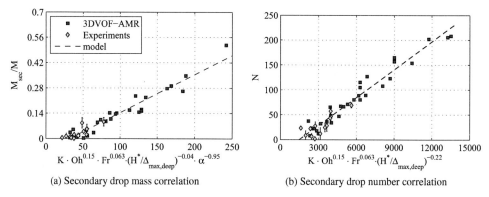

(a) Secondary drop mass correlation   (b) Secondary drop number correlation

Fig. 7.5: Graphical verification of the correlations for secondary drop mass and number during single drop impingement onto liquid surfaces

Secondary Drop Sizes and Velocities

In chapter 5 and 6, it has already been demonstrated that the secondary drop size distribution can be described by means of log-normal density distribution functions. Here, the required mean parameter corresponds to the arithmetic mean secondary drop size D_{10}/D. The related spread factor representing the width of the distribution function is determined with the logarithmic standard deviation $\sigma_{ln,D}$. The log-normal probability distribution function is defined as

$$f_0(D) = \frac{1}{D\sigma_{ln,D}\sqrt{2\pi}}exp\left\{-\frac{1}{2}\left[\frac{ln(D/D_{10})}{\sigma_{ln,D}}\right]^2\right\}, \tag{7.4}$$

where D is the secondary drop diameter, $\sigma_{ln,D}$ the width of the distribution and D_{10} the arithmetic mean diameter.

The correlations based on equation 7.2 for the arithmetic mean secondary drop diameter and the logarithmic standard deviation are presented in table 7.5. Their verification by comparison with the experimental and numerical results presented in Chapter 5 and 6 can be found in the Figs 7.6a and 7.6b.

For the secondary drop velocities, correlations were only developed for the arithmetic mean parallel and orthogonal velocity component. The appropriate constants are listed in table 7.5. The correlations are compared to the experimental and numerical results in Figures 7.6c and 7.6d. Here, in contrast to the previous correlations, a large scattering of the data occurs. Hence, the proposed correlation may give only an approximate value of the secondary drop velocities. However, in most technical applications the air flow which is present in direct vicinity of the impingement location is known to be much faster than the ejected secondary drop velocity. The secondary droplets are hence subject to strong accelerations; their initial ejection velocity is hence not decisive with respect to the further droplet trajectories.

P	a	b	c	d	e	f	g
D_{10}/D	0.3015	1	-0.6	0	0	-3	0.069894
U_D	0	±0.05	±0.1	0	0	±0.1	0
$\sigma_{ln,D}$	0.0048535	1	-0.4	0.05	0	-1.25	0.15585
U_σ	0	±0.05	±0.1	±0.02	0	±0.1	0
$V_{mean,pll,sec}/V$	1788	-1.0292	0.288	0	0.10292	0	0
U_{Vpll}	0	±0.102	±0.102		±0.051	0	0
$V_{mean,orth,sec}/V$	0.0063708	1	0	0	0	-1	0.45068
U_{Vorth}	0	±0.05	0	0	0	±0.1	0

Table 7.5: Coefficients and exponents for the secondary drop sizes and velocities during single drop impingement onto liquid surfaces including the uncertainty U_x in the powers

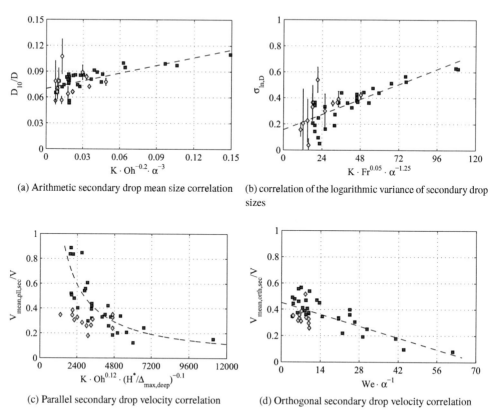

(a) Arithmetic secondary drop mean size correlation

(b) correlation of the logarithmic variance of secondary drop sizes

(c) Parallel secondary drop velocity correlation

(d) Orthogonal secondary drop velocity correlation

Fig. 7.6: Graphical verification of the correlations for secondary drop sizes and velocities during single drop impingement on liquid surfaces

7.4   Drop-Film Interaction Model

Beside the fundamental characterization of the splashing products during drop impingement with and without the wall-effects, one of the main targets in the present thesis was the development of a drop-film interaction model applicable for the range of impingement conditions found in oil system flows of aero-engines. The combination of the correlations developed in the previous subsection and the non-splash/splash threshold described in chapter 2 allow hence the complete formulation of a drop-film interaction model that is described in detail in the appendix. The validity of the drop-film interaction model for impingement conditions found in bearing chambers is confirmed by Fig. 7.7. Here, a comparison between the range of impingement conditions found in bearing chambers and the investigated impingements in this thesis is shown. In particular, the

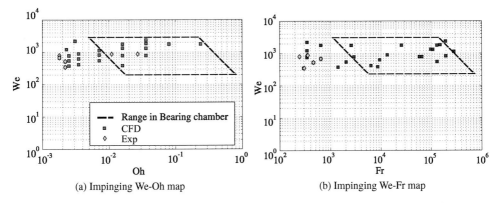

(a) Impinging We-Oh map   (b) Impinging We-Fr map

Fig. 7.7: Comparison of the investigated experimental and numerical range of single drop impingement events with the impingement conditions derived from measurements in the bearing chamber

impingement with large Ohnesorge numbers and Froude numbers as they occur very frequently in bearing chambers are well covered by the drop-film interaction model. The complete range of validity is furthermore suggested below.

- $300 < We < 2376$

- $0.0018 < Oh < 0.047$

- $136 < Fr < 282928$

- $0.1 < H^*/\Delta_{max,deep} < 1$

- $30° < \alpha < 90°$

7.5   Significance of the Model for Numerical Simulation of Aero-Engine Oil Systems

The drop-film interaction model can be well included in Euler-Lagrange CFD Simulations of the bearing chamber two-phase flow. On the one hand, it allows a simple coupling with the available Lagrangian Particle Tracking approach often used for the description of the droplet propagation in bearing chambers (see e.g. Glahn et al. (1997), Aggarwal and Peng (1995), Weinstock and Heister (1997), Farral (2000), Simmons et al. (2002), Weinstock and Heister (2006), Aidarinis et al. (2010), and Hohenreuther (2011)). On the other hand, good coupling of the drop-film interaction model with codes describing the motion of the liquid wall film flow is also possible. For simulation of the thin wall film mostly occurring near the vent-off-takes in bearing chamber, these models are usually referred to as eulerian thin film models (see i.e. Farrall et al. (2003), Farrall et al. (2006), Weinstock and Heister (2006), O'Rourke and Amsden (2000)). The numerical description of thick liquid wall films on the other hand as they occur near the scavenge off-takes in bearing chamber require a more detailed resolution of the film dynamics (i.e. see the VOF method in Peduto (2009), Peduto et al. (2011a) and Hashmi (2012)).

Hence, the integration of the drop-film interaction model paves the way for enhanced prediction capability not only for simulation methods in oil system flows of aero-engines but also in other practical applications with similar parameter ranges. The bearing chamber two-phase flow, specifically, can be determined in more detail. By improving the prediction of the secondary spray generation arising from drop-film interaction, more accurate boundary conditions can be defined for a more robust design of the labyrinth seals with respect to oil leakage. Both will contribute to lower oil consumption, improvements in cabin air contamination, and resistance against oil fire. Simultaneously, the design of air-oil separators and breathers in the venting pipe are also enhanced with a better prediction of the spray characteristics in bearing chambers. Beside the sprays, moreover, the mass and momentum exchange in the liquid wall films due to drop impingement in bearing chambers can also be evaluated with more accuracy. This contributes at the same time to an enhanced evaluation of the oil cooling performance of the surrounding hot compartments as one of the major functions of the oil in the rear bearing chambers.

8   Summary and Outlook

The trend in the present and future aviation market towards higher power outputs, increased thermal and fuel efficiency, and acceptable environmental impact of aero-engines, has made vital the role of the design of the heat management and oil systems in future aero-engines. Within the oil flow system, the design of a reliable, robust and efficient bearing chamber is considered to be the most challenging part, due to the complexity of the coincident two-phase flow phenomena, namely the wall films attached to the chamber walls, the droplet laden air flows, and the drop-film interactions. A special feature of these flow phenomena in bearing chambers is their strong dependence on the geometric configuration and engine operation cycle. This makes the transfer of empirical correlations derived from experimental models and generic test rigs questionable. The growing potential of the multiphase CFD technique, together with increasing computational resources, has paved the way for the development of an integrated CFD multiphase modelling technique. Such a simulation tool should be capable of calculating the complete two-phase flows and heat transfer in oil systems. Despite the significant progress over the last decade in the modelling of the droplet and wall film dynamics in bearing chambers, there has still been a lack of a physical sub-model accounting for the description of the mass and momentum exchange between the impinging drops and the wall films. Drop-film interaction is known to affect the overall performance of an oil-system. The sprays that are generated through the impingement of drops onto wall films contain very small drop sizes, that may leak out of the bearing chambers, causing cabin air contamination and serious safety issues through eventual oil fires. Moreover, the cooling performance of the wall film flowing attached to the wall in bearing chambers is also decisively influenced by the impinging drops. For this reason, a detailed understanding and modelling of the drop-to-film interaction is inevitable for the future simulation tools of oil system flow design.

The main aim of the present investigation was to develop a physical understanding and a drop-film interaction model for the parameter range found in the bearing chambers of aero-engines. The mass and momentum exchange between droplets and liquid wall films was characterised by analysing the secondary droplets generated during single droplet impingement onto liquid wall films. However, the small drop sizes together with their large impingement velocity, as occurring in these bearing chambers, challenged the direct experimental measurement of the splashing products for the appropriate scales of the impingement times and lengths. With the help of an enhanced numerical CFD method based on the Volume-Of-Fluid method, and coupled with an Adaptive Mesh Refinement technique, direct simulation of the secondary drop generation during a splashing impingement was achieved. Two different numerical setups were developed and applied in the present investigation. For the prediction of the cavity penetration and the lamella expansion during single drop impingements, an axisymmetric 2D-VOF-AMR method was used. The secondary drop generation due to the corona break-up is a fully three-dimensional phenomenon, and therefore required the 3D-VOF-AMR method. In order to prove the capability of both numerical setups, a thorough experimental validation of these numerical methods was conducted.

Here, the 2D-VOF-AMR method was validated using measurements of the cavity penetration during millimetre size single drop impingement onto deep liquid pools from the literature. The validation of the 3D-VOF-AMR method was accomplished using experimental results derived from our own experimental measurements on millimetre size drop impingements. The measurements were carried out using a generic test facility and an enhanced Shadowgraphy and Particle Tracking Velocimetry technique developed at the Institute for Thermal Turbomachinery at the KIT. The experimental validation of both numerical setups showed very good correspondence for the cavity and lamella evolution as well as the secondary drop characteristics. It proved to work robustly and with considerably lower computational effort compared to other numerical methodologies published in the literature. With the validated numerical methods, hence, a database could be generated with the help of various single drop impingement events from which correlations for the drop-film interaction model could be derived. The parametric study of the secondary drop generation during single drop impingement was divided into two parts: one neglecting, and the other including, the wall effects coming from the bottom wall. Different factors influencing the secondary drop generation, such as the Weber number, the Ohnesorge number, the Froude number, and the impact angle, have been analysed in detail. One of the major findings of this study was the role of gravitational or centrifugal forces in the drop impact dynamics. It showed on the one hand that a change in gravitational or centrifugal forces within an impingement can considerably change the cavity penetration, lamella expansion, and secondary drop generation. For the cavity and crown evolution, for instance, an increase of the Froude number by reducing the gravitational/centrifugal acceleration resulted in much larger maximum cavity depth and crown heights. For the secondary drop generation, it was shown that an increase in the Froude number, by reducing the impinging droplet size and maintaining the drop momentum constant, caused moreover an enhanced number of secondary drops produced compared to the large millimetre size drop impingements. These new findings proved that coarse spray impingement models derived from single drop impingement data and applying only a non-dimensional momentum for scaling, as e.g. the K number $K = WeOh^{-0.4}$, are inaccurate and do not take into account the important influence of the gravitational/centrifugal forces. The inclusion of the Froude number in correlations describing the mass and momentum exchange in drop-film interaction models is unavoidable. However, none of the drop-film interaction models available in the literature take into account the influence of this parameter, which is significant. Another important achievement of this thesis was the improved characterization of the wall-effects influencing the secondary drop generation. The presented numerical results confirmed the findings of Bisighini (2010). An effect on the secondary drop generation arising from the bottom wall could only be observed if the cavity penetrates down to and interacts with the bottom wall. Therefore, the ratio $\Delta_{max,deep}/H^*$ proved to be the determining parameter describing the wall-effects. The relative film height H^* only is not sufficient for description of the wall-effects. With respect to the effect of the impingement angle, an interesting outcome could also be extracted from the numerical results. The secondary drop characteristics and ejected mass fraction followed by oblique impingement onto liquid surfaces differed significantly from the normal impingements. The effect of the impact angle showed a strong increase of the mean secondary drop size and mass from reducing the angle from $\alpha = 90°$ to $\alpha = 45°$ whereas reducing again for even flatter impact angles.

The cavity penetration in oblique impingements was proportional to the normal impingement velocity component. The crown expansion, in contrast, is more complex and follows a more non-linear behaviour. Beside the advances in fundamental physics describing the drop impact dynamics with the help of the numerical and experimental results, a set of correlations was also derived, creating the drop-film interaction model aimed. The validity of the correlations could be claimed since all of the numerical results were derived from drop impingement events within the parameter range found in bearing chambers. By integrating the present sub-model into CFD multiphase simulation methods for aero-engine oil flow system design, new horizons are opened for the prediction of the oil sealing and cooling performance of bearing chambers.

References

Agbaglah, G. and Deegan, R.D. (2014): *Growth and Instability of the Liquid Rim in the Crown Splash Regime*. Journal of Fluid Mechanics, Vol. 752, pp 485–496.

Agbaglah, G., Josserand, C. and Zaleski, S. (2013): *Longitudinal Instability of a Liquid Rim*. Physics of Fluids, Vol. 25, 102104.

Aggarwal, S.K. and Peng, F. (1995): *A Review of Droplet Dynamics and Vaporization Modeling for Engineering Calculations*. Journal of Engineering for Gas Turbines and Power, Vol. 117, pp 453–461.

Aidarinis, J., Missirlis, D., Yakinthos, K. and Goulas, A. (2010): *CFD Modelling and LDA Measurements for the Air-flow in an Aero-Engine Front Bearing Chamber*. ASME Conference Proceedings, Vol. GT2010-23294.

Alghoul, S., Eastwick, C. and Hann, D (2011): *Normal Droplet Impact on Horizontal Moving Films: An Investigation of Impact Behaviour and Regimes*. Experiments in Fluids, Vol. 50, pp 1305–1316.

Babinsky, E. and Sojka, P.E. (2002): *Modeling Drop Size Distributions*. Progress in Energy and Combustion Science, Vol. 28, pp 303–329.

Bai, C. and Gosman, A.D. (1995): *Development of Methodology for Spray Impingement Simulation*. SAE Conference Proceedings, Vol. 5, pp 1249–1257.

Barth, J. and Jespersen, D. (1989): *The Design and Application of Upwind Schemes on Unstructured Meshes*. AIAA 27th Aerospace Sciences Meeting, Reno, Nevada, Vol. Technical Report.

Berberović, E., van Hinsberg, N., Jakirlić, S. and Roisman, I.and Tropea, C. (2009): *Drop Impact onto a Liquid Layer of Finite Thickness: Dynamics of the Cavity Evolution*. Phys. Rev. E, Vol. 79, pp 36–46.

Birkenkaemper, S. (1996): *Untersuchung der Zweiphasenströmung in TLK und deren Entlüftungssystem*. Diplomarbeit, Institute for Thermal Turbomachinery, KIT.

Bisighini, Alfio (2010): *Single and Double Drop Impacts onto Deep and Thick Liquid Layers*. PhD Thesis, University of Bergamo.

Bisighini, Alfio, Cossali, Gianpietro E., Tropea, Cameron and Roisman, Ilia V. (2010): *Crater Evolution after the Impact of a Drop onto a Semi-Infinite Liquid Target*. Phys. Rev. E, Vol. 82,036319.

Brackbill, J.U, Kothe, D.B and Zemach, C (1992): *A Continuum Method for Modeling Surface Tension*. Journal of Computational Physics, Vol. 100, pp 335 – 354.

Brambilla, P. and Guardone, A. (2013): *Automatic Tracking of Corona Propagation in Three-Dimensional Simulations of Non-Normal Drop Impact on a Liquid Film*. Journal of Computing, Vol. 95.

Bremond, N. and Villermaux, E. (2006): *Atomization by Jet Impact*. Journal of Fluid Mechanics, Vol. null, pp 273–306.

Brutin, D. (2003): *Drop Impingement on a Deep Liquid Surface: Study of a Crater's Sinking Dynamics*. Comptes Rendus Mecanique, Vol. 331, pp 61 – 67.

Busam, S., Glahn, A. and Wittig, S. (2000): *Internal Bearing Chamber Wall Heat Transfer as a Function of Operating Conditions and Chamber Geometry*. Journal of Engineering for Gas Turbines and Power, Vol. 122, pp 314–320.

Bussmann, M., Mostaghimi, J. and Chandra, S. (1999): *On a Three-Dimensional Volume Tracking Model of Droplet Impact*. Physics of Fluids, Vol. 11, pp 1406–1417.

Bussmann, M., Chandra, S. and Mostaghimi, J. (2000): *Modeling the Splash of a Droplet Impacting a Solid Surface*. Physics of Fluids, Vol. 12, pp 3121–3132.

Chandra, B., Simmons, K., Pickering, S. and Tittel, M. (2010): *Factors Affecting Oil Removal From an Aeroengine Bearing Chamber*. ASME Conference Proceedings, Vol. 2010, pp 219–228.

Coghe, A., Cossali, G.E. and Marengo, M. (1995): *A First Study about Single Drop Impingement on a Thin Liquid Film in a Low Laplace Number Range*. ICLASS, Nuernberg, Germany.

Cossali, G., Coghe, A. and Marengo, M. (1997): *The Impact of a Single Drop on a Wetted Solid Surface*. Experiments in Fluids, Vol. 22, pp 463–472.

Cossali, G.E., Brunello, G., Coghe, A. and Marengo, M. (1999): *Impact of a Single Drop on a Liquid Film: Experimental Analysis and Comarison with Empirical Models*. Italian Congress of Thermal Dynamics UIT, Ferrara.

Cossali, G.E., Marengo, M., Coghe, A. and Zhdanov, S. (2004): *The Role of Time in Single Drop Splash on Thin Film*. Experiments in Fluids, Vol. 36, pp 888–900.

Cossali, G.E., Marengo, M. and Santini, M. (2005): *Single Drop Empirical Models for Spray Impact on Solid Walls: A Review*. Atomization and Sprays, Vol. 15.

Da Vinci, L. (1508): *The Notebooks of Leonardo da Vinci*. ed. and trans. E. MacCurdy. George Brazilier, Codex Leicester.

Daunenhofer, J.F. and Baron, J.R. (1985): *Grid Adaption for the 2D Euler Equations*. Technical Report AIAA.

Dullenkopf, K., Willmann, M., Wittig, S., Schoene, F., Stieglmeier, M., Tropea, C. and Mundo, C. (1998): *Comparative Mass Flux Measurements in Sprays using a Patternator and the Phase-Doppler Technique*. Particle and Particle Systems Characterization, Vol. 15, pp 81–89.

Ebner, J. (2004): *Einfluss von Druckgradienten in der Gasströmung auf die Dynamik schubspannungsgetriebener Wandfilme*. PhD Thesis, Institute for Thermal Turbomachinery, University of Karlsruhe.

Ebner, J., Gerendás, M., Schäfer, O. and Wittig, S. (2002): *Droplet Entrainment From a Shear-Driven Liquid Wall Film in Inclined Ducts: Experimental Study and Correlation Comparison*. Journal of Engineering for Gas Turbines and Power, Vol. 124, pp 874–880.

Ebner, J., Schober, P., Schaefer, O., Koch, R. and Wittig, S. (2004): *Modelling of Shear Driven Liquid Wall Films: Effect of Accelerated Air flow on the Film Flow Propagation*. Progress in Computational Fluid Dynamics, Vol. 4, pp 183–190.

Eggers, J., Fontelos, M., Josserand, C. and Zaleski, S. (2010): *Drop Dynamics after Impact on a Solid Wall: Theory and Simulations*. Physics of Fluids, Vol. 22, pp 62–76.

Elsaesser, A. (1998): *Kraftstoffausbreitung in Verbennungskraftmaschinen: Grundlagen der Strömungs schubspannungsgetriebener Wandfilme*. PhD Thesis, Institute for Thermal Turbomachinery, University of Karlsruhe.

Engel, Olive G (1966): *Crater Depth in Fluid Impacts*. Journal of Applied Physics, Vol. 37, pp 1798–1808.

Farral, M. (2000): *Numerical Modelling of Two-phase Flow in a Simplified Bearing Chamber*. PhD Thesis, University of Nottingham.

Farrall, M., Hibberd, S. and Simmons, K. (2003): *Modeling Oil Droplet/Film Interaction in an Aero-Engine Bearing Chamber*. ICLASS - 2003 <Sorrento, Italy, 2003>.

Farrall, M., Simmons, K., Hibberd, S. and Gorse, P. (2006): *A Numerical Model for Oil Film Flow in an Aeroengine Bearing Chamber and Comparison to Experimental Data*. Journal of Engineering for Gas Turbines and Power, Vol. 128, pp 111–117.

Fedorchenko, A.I. and Wang, A. (2004): *On Some common Features of Drop Impact on Liquid Surfaces*. Physics of Fluids, Vol. 16, pp 1349–1365.

Ferziger, J. and Peric, M. (2008): In: Springer (Hrsg.), *Computational Fluid Dynamics*, S. 509.

Fest-Santini, S., Guilizzoni, M., Santini, M. and Cossali, G.E. (2011): *Drop Impacts in Pools: A Comparison between High-Speed Imaging and Numerical Simulations*. DIPSI Workshop 2011, Bergamo, Italy.

Fest-Santini, S., Guilizzoni, M., Santini, M. and Cossali, G.E. (2012): *Water Drop Impact into a Deep Pool: Influence of the Liquid Pool Temperature*. DIPSI Workshop 2012, Bergamo, Italy.

Flouros, M. (2005): *The Impact of Oil and Sealing Air Flow, Chamber Pressure, Rotor Speed, and Axial Load on the Power Consumption in an Aeroengine Bearing Chamber*. Journal of Engineering for Gas Turbines and Power, Vol. 127, pp 182–186.

Flouros, M. (2006): *Reduction of Power Losses in Bearing Chambers Using Porous Screens Surrounding a Ball Bearing*. Journal of Engineering for Gas Turbines and Power, Vol. 128, pp 178–182.

Fullana, J. and Zaleski, S. (1999): *Stability of a Growing End Rim in a Liquid Sheet of Uniform Thickness*. Physics of Fluids, Vol. 11, pp 952–954.

Fuster, D., Bague, A., Boeck, T., Le Moyne, L., Leboissetier, A., Popinet, S., Ray, P., Scardovelli, R. and Zaleski, S. (2009): *Simulation of Primary Atomization with an Octree Adaptive Mesh Refinement and VOF Method*. International Journal of Multiphase Flow, Vol. 35, pp 550 – 565.

Galusinski, C. and Vigneaux, P. (2008): *On Stability Condition for Bifluid Flows with Surface Tension: Application to Microfluidics*. J. Comput. Phys., Vol. 227, pp 6140–6164.

Geldorp, W.I., Rioboo, R., Jakirlic, S., Muzaferija, S. and Tropea, C. (2000): *Numerical and Experimental Drop Impact on Solid Dry Surfaces*. ICLASS 2000, Pasadena, CA, USA.

Gepperth, S., Guildenbecher, D., Koch, R. and Bauer, H.-J. (2010): *Pre-filming Primary Atomization: Experiments and Modeling*. ILASS-Europe 2010.

Gingold, R.A. and Monaghan, J.J. (1977): *Smoothed Particle Hydrodynamics: Theory and Application to Non-Spherical Stars*. Monthly notices of the Royal Astronomical Society, Vol. 181, pp 375–389.

Glahn, A. (1995): *Zweiphasenstroemung in Triebwerkslagerkammern - Charakterisierung der Oelfilmstroemung und des Waermeuebergangs*. PhD Thesis, Institut fuer Thermische Stroemungsmaschinen, Karlsruher Institut fuer Technologie.

Glahn, A. and Wittig, S. (1996): *Two-Phase Air/Oil Flow in Aero Engine Bearing Chambers: Characterization of Oil Film Flows*. Journal of Engineering for Gas Turbines and Power, Vol. 118, pp 578–583.

Glahn, A., Kurreck, M., Willmann, M. and Wittig, S. (1997): *Feasibility Study on Oil Droplet Flow Investigations inside Aero-Engine Bearing Chambers-PDPA Techniques in Combination with Numerical Approaches*. Journal of Engineering for Gas Turbine and Power, Vol. 118, pp 749–755.

Glahn, A., Busam, S., Blair, M. F., Allard, K. L. and Wittig, S. (2002): *Droplet Generation by Disintegration of Oil Films at the Rim of a Rotating Disk*. Journal of Engineering for Gas Turbines and Power, Vol. 124, pp 117–124.

Glahn, A., Blair, M. F., Allard, K. L., Busam, S., Schäfer, O. and Wittig, S. (2003): *Disintegration of Oil Films Emerging From Radial Holes in a Rotating Cylinder*. Journal of Engineering for Gas Turbines and Power, Vol. 125, pp 1011–1020.

Gloeckner, P. (2011): *Zuverlaessigkeit- und Effizienzsteigerung von Triebwerkskugellagern*. PhD Thesis, Institute for Thermal Turbomachinery, KIT Karlsruhe.

Gomaa, H. and Weigand, B. (2012): *Modelling and Investigation of the Interaction between Drops and Blades in Compressor Cascades with a Droplet Laden Inflow*. ISROMAC-14.

Gomaa, H., Weigand, B., Haas, M. and Munz, C.-D. (2009): *Direct Numerical Simulation (DNS) on the Influence of Grid Refinement for the Process of Splashing*. In: Nagel, W., Kröner, D. and Resch, M. (Hrsg.), *High Performance Computing in Science and Engineering '08*, pp 241–255.

Gomaa, H., Stotz, I., Sievers, M., Lamanna, G. and Weigand, B. (2011): *Preliminary Investigation on Diesel Droplet Impact on Oil Wallfilms in Diesel Engines*. ILASS Europe - 2011 <Estoril, Portugal,2011>.

Gorse, P. (2007): *Tropfenentstehung und Impulsaustausch in Lagerkammern von Flugtriebwerken*. PhD Thesis, Institute for Thermal Turbomachniery, KIT.

Gorse, P., Willenborg, K., Busam, S., Ebner, J., Dullenkopf, K. and Wittig, S. (2003): *3D Laser Doppler Anemometer Measurements in Aer-Engine Bearing Chamber*. ASME Conference Proceedings: GT2003-38376.

Gorse, P., Busam, S. and Dullenkopf, K. (2004): *Influence of Operating Condition and Geometry on the Oil Film Thickness in Aero-Engine Bearing Chambers*. ASME Conference Proceedings: GT2004-41693.

Gorse, P., Dullenkopf, K., Bauer, H.-J. and Wittig, S. (2008): *An Experimental Study on Droplet Generation in Bearing Chambers Caused by Roller Bearings*. ASME Conference Proceedings, Vol. 2008, pp 1681–1692.

Gueyffier, D. and Zaleski, S. (1998): *Formation de digitations lors de l'impact d'une goutte sur un film liquide*. Comptes Rendus de l'Academie des Sciences - Series IIB - Mechanics-Physics-Astronomy, Vol. 326, pp 839 – 844.

Han, Z., Xu, Z. and Trigui, N. (2000): *Spray/Wall Interaction Models for Multidimensional Engine Simulation*. International Journal of Engine Research, Vol. 1, pp 127–146.

Harvie, D.J.E., Davidson, M.R. and Rudman, M. (2006): *An Analysis of Parasitic Current Generation in Volume of Fluid simulations*. Applied Mathematical Modelling, Vol. 30, pp 1056 – 1066.

Hashmi, A. A. (2012): *Oil film dynamics in Aero Engine Bearing Chambers - Fundamental Investigations and Numerical Modelling*. PhD Thesis, Institute for Thermal Turbomachinery, KIT.

Hecht, E. (2002): *Optik*. Oldenburg, Munich.

Himmelsbach, J. (1992): *Zweiphasenstroemungen mit schubspannungsgetriebenen welligen Flüssigkeitsfilmen in turbulenter Heissluftströmung*. PhD Thesis, Institute for Thermal Turbomachinery, University of Karlsruhe.

Himmelsbach, J., Noll, B. and Wittig, S. (1994): *Experimental and Numerical Studies of Evaporating Wavy Fuel Films in Turbulent Air Flow*. International Journal of Heat and Mass Transfer, Vol. 37, pp 1217 – 1226.

Hirt, C.W. and Nichols, B.D. (1981): *Volume of Fluid (VOF) Method for the Dynamics of Free Boundaries*. Journal of Computational Physics, Vol. 39, pp 201–225.

Hoefler, C., Braun, S., Koch, R. and Bauer, H.-J. (2012): *Multiphase Flow Simulations Using the Meshfree Smoothed Particle Hydrodynamics Method*. ICLASS - 2012 <Heidelberg, Germany, 2012>.

Hoefler, C., Braun, S., Koch, R. and Bauer, H.-J. (2013): *Modeling Spray Formation in Gas Turbines: A New Meshless Approach*. Journal of Engineering for Gas Turbines and Power, Vol. 135, pp 453–461.

Hohenreuther, U. (2011): *Entwicklung eines numerischen (RANS) Modells zur Beschreibung der Gasströmung in Triebwerkslagerkammern*. Studienarbeit, Institute for Thermal Turbomachinery, KIT.

Huser, A. (2011): *Experimentelle Untersuchung der Destruktionsgrenze bei der Tropfen-Wandfilm Interaktion unter lagerkammerähnlichen Interaktionsbedingungen*. Studienarbeit, Institute for Thermal Turbomachniery, KIT.

Issa, R.I. (1986): *Solution of the Implicitly Discretized Fluid Flow Equations by Operator-Splitting*. J. Comput. Phys., Vol. 62, pp 40–65.

Jayaratne, O. W. and Mason, B. J. (1964): *The Coalescence and Bouncing of Water Drops at an Air/Water Interface*. Proceedings of the Royal Society of London. Series A. Mathematical and Physical Sciences, Vol. 280, pp 545–565.

Josserand, J. and Zaleski, S. (2003): *Droplet Splashing on a Thin Liquid Film*. Physics of Fluids, Vol. 15, pp 1650–1657.

Kalantari, Davood (2007): *Characterization of Liquid Spray Impact onto Walls and Films*. PhD Thesis, TU Darmstadt.

Kapulla, R., Tuchtenhagen, J., Mueller, A., Dullenkopf, K. and Bauer, H.-J. (2008): *Droplet Sizing Performance of Different Shadow Sizing Codes*. Gesellschaft fuer Laser-Anemometrie - GALA, Karlsruhe, Germany.

Kline, S.J. and McClintock, F.A. (1953): *Describing Uncertainties in Single-Sample Experiments*. Mech. Eng., p. 3.

Kontin, S (2015): *to be finished: Einsatz von Harnstoffwasserloesung zur Abgasnachbehandlung*. PhD Thesis, Institut fuer Thermische Stroemungsmaschinen, Karlsruher Institut fuer Technologie.

Krechetnikov, R. and Homsy, G.M. (2009): *Crown-forming Instability Phenomena in the Drop Splash Problem*. Journal of Colloid Interface Science, Vol. 331.

Krug, M. B., Peduto, D., Kurz, W. and Bauer, H.-J. (2015): *Experimental Investigation Into the Efficiency of an Aero-engine Oil Jet Supply System*. ASME Journal of Engineering for Gas Turbines and Power, Vol. 137,011505, pp 1–7.

Kruisbrink, A., Pearce, F., Yue, T., Cliffe, K. and Morvan, H. (2011): *SPH Concepts for Continous Wall and Pressure boundaries*. SPHERIC Workshop.

Kurz, W., Dullenkopf, K. and Bauer, H.-J. (2012): *Influences on the Oil Split Between the Offtakes of an Aero-Engine Bearing Chamber*. ASME Conference Proceedings, Vol. GT2012-69412.

Kyriopoulos, O. (2010): *Gravity Effect on Liquid Film Hydrodynamics and Spray Cooling*. PhD Thesis, Strömungslehre und Aerodynamik, FB Maschinenbau, TU Darmstadt.

Lee, C.W., Palma, P.C. Simmons, K. and Pickering, S.J. (2005): *Comparison of CFD and PIV Data for the Airflow in an Aero-engine Bearing Chamber*. Journal of Engineering for Gas Turbines and Power, Vol. 127, pp 697–705.

Leneweit, G., Koehler, R., Roesner, K. G. and Schaefer, G. (2005): *Regimes of Drop Morphology in Oblique Impact on Deep Fluids*. Journal of Fluid Mechanics, Vol. 543, pp 303–331.

Levin, Z. and Hobbs, P. V. (1971): *Splashing of Water Drops on Solid and Wetted Surfaces: Hydrodynamics and Charge Separation*. Vol. 269, pp 555–585.

Liow, J.L. (2001): *Splash Formation by Spherical Drops*. Journal of Fluid Mechanics, Vol. 427, pp 73–105.

Macklin, W. C. and Metaxas, G. J. (1976): *Splashing of Drops on Liquid Layers*. Journal of Applied Physics, Vol. 47, pp 3963 –3970.

Mandre, S. and Brenner, M. (2012): *The Mechanism of a Splash on a Dry Solid Surface*. Journal of Fluid Mechanics, Vol. 690, pp 148–172.

Mathews, H.-C., Morvan, H., Peduto, D., Wang, Y., Young, C. and Bauer, H.-J. (2013): *Modelling of Hydraulic Seals using an Axisymmetric Volume Of Fluid Method*. ASME Conference Proceedings: GT2013-95070.

Morton, D., Rudman, M. and Liow, J.-L. (2000): *An Investigation of the Flow Regimes Resulting from Splashing Drops*. Physics of Fluids, Vol. 12, pp 747–763.

Muehlbauer, M., Roisman, I. and Tropea, C. (2010): *Evaluation of Spray/Wall Interaction Data*. 15th Int Symp on Appl of Laser Techniques to Fluid Mechanics.

Mueller, A., Koch, R., Bauer, H.-J., Hehle, M. and Schaefer, O. (2006): *Performance of Prefilming Airblast Atomizers in Unsteady Flow Conditions*. ASME Conference Proceedings: GT2006-90432.

Mueller, A., Dullenkopf, K. and Bauer, H.-J. (2008): *Application of an Extended Particle Tracking Method to Analyze Droplet Wall Interaction.* 14th Int Symp on Appl of Laser Techniques to Fluid Mechanics.

Mueller, A., Velios, E., Dullenkopf, K. and Bauer, H.-J. (2009): *From Single Droplet to Spray Wall Interaction - Multiple Droplet Chains.* ICLASS - 2009 <Vail, Colorado USA, 2009>.

Mugele, R. A. and Evans, H. D. (1951): *Droplet Size Distribution in Sprays.* Industrial and Engineering Chemistry, Vol. 43, pp 1317–1324.

Mundo, Chr., Sommerfeld, M. and Tropea, C. (1995): *Droplet-Wall Collisions: Experimental Studies of the Deformation and Breakup Process.* International Journal of Multiphase Flow, Vol. 21, pp 151 – 173.

Mundo, C., Tropea, C. and Sommerfeld, M. (1997): *Numerical and Experimental Investigation of Spray Characteristics in the Vicinity of a Rigid Wall.* Experimental Thermal and Fluid Science, Vol. 15, pp 228 – 237.

Nikolopoulos, N., Theodorakakos, A. and Bergeles, G. (2007): *Three-Dimensional Numerical Investigation of a Droplet Impinging Normally onto a Wall-Film.* Journal of Computational Physics, Vol. 225, pp 322 – 341.

Okawa, T., Shiraishi, T. and Mori, T. (2006): *Production of Secondary Drops during the Single Water Drop Impact onto a Plane Water Surface.* Experiments in Fluids, Vol. 41, pp 965–974.

Okawa, T., Shiraishi, T. and Mori, T. (2008): *Effect of Impingement Angle on the Outcome of Single Water Drop Impact onto a Plane Water Surface.* Experiments in Fluids, Vol. 44, pp 331–339.

O'Rourke, P.J. and Amsden, A.A. (2000): *A Spray/Wall Interaction Submodel for the KIVA-3 Wall Film Model.* SAE Technical Paper 2000-01-0271 - 2000 <, Detroit, Michigan, United States.

Pasandideh-Fard, M., Qiao, Y. M., Chandra, S. and Mostaghimi, J. (1996): *Capillary Effects during Droplet Impact on a Solid Surface.* Physics of Fluids, Vol. 8, pp 650–659.

Peduto, D. (2009): *CFD Modelling of Simplified Tangential Scavenge Off-take flows in Aero-Engine Bearing Chambers.* Diplomarbeit, Institut für Thermische Strömungsmaschinen, KIT.

Peduto, D., Hashmi, A., Dullenkopf, K., Bauer, H.-J. and Morvan, H. (2011a): *Modelling of an Aero-Engine Bearing Chamber Using Enhanced CFD Technique.* ASME Conference Proceedings, Vol. 2011, pp 809–819.

Peduto, D., Koch, R., Morvan, H. and Bauer, H.-J. (2011b): *Numerical Studies of Single Drop Impact onto a Shallow and Deep Liquid Pool.* ILASS-Europe 2011, Estoril, Portugal.

Plateau, J. (1873): *Statique Experimentale et Theorique des Liquides Soumis aux Seules Forces Moleculaires*. Gauthier Villars, Vol. 2.

Popinet, S. (2009): *An Accurate Adaptive Solver For Surface-Tension-Driven Interfacial Flows*. Journal of Computational Physics, Vol. 228, pp 5838 – 5866.

Rayleigh, Lord (1878): *On The Instability Of Jets*. Proceedings of the London Mathematical Society, Vol. s1-10, pp 4–13.

Rayleigh, Lord (1892): *On the Instability of a Cylinder of Viscous Liquid under Capillary Force*. Phil. Mag., Vol. 34, pp 145–154.

Rein, M. and Delplanque, J.-P. (2008): *The Role of Air Entrainment on the Outcome of Drop Impact on a Solid Surface*. Acta Mechanica, Vol. 201, pp 105–118.

Richter, B., Dullenkopf, K. and Bauer, H.-J. (2005): *Investigation of Secondary Droplet Characteristics produced by an Isooctane Drop Chain Impact onto a Heated Piston Surface*. Experiments in Fluids, Vol. 39, pp 351–363.

Richtmyer, R. D. (1960): *Taylor Instability in Shock Acceleration of Compressible Fluids*. Communications on Pure and Applied Mathematics, Vol. 13, pp 297–319.

Rider, W. and Kothe, D. (1998): *Reconstructing Volume Tracking*. Computational Physics, Vol. 42, pp 112–142.

Rieber, M. and Frohn, A. (1999): *A Numerical Study on the Mechanism of Splashing*. International Journal of Heat and Fluid Flow, Vol. 20, pp 455 – 461.

Rioboo, R., Marengo, M. and Tropea, C. (2002): *Time Evolution of Liquid Drop Impact onto Solid, Dry Surfaces*. Experiments in Fluids, Vol. 33, pp 112–124.

Rioboo, R., Bauthier, C., Conti, J., Voue, M. and De Coninck, J. (2003): *Experimental Investigation of Splash and Crown Formation during Single Drop Impact on Wetted Surfaces*. Experiments in Fluids, Vol. 35, pp 648–652.

Robinson, A. Morvan, H. and Eastwick, C. (2010): *Computational Investigations Into Draining in an Axisymmetric Vessel*. Journal of Fluids Engineering, Vol. 132, pp 104–121.

Rodriguez, F. and Mesler, R. (1988): *The Penetration of Drop-Formed Vortex Rings into Pools of Liquid*. Journal of Colloid and Interface Science, Vol. 12, pp 121 – 129.

Roisman, I. and Tropea, C. (2001): *Flux Measurements in Sprays using Phase Doppler Techniques*. Atomization and Sprays, Vol. 11, pp 667–699.

Roisman, Ilia V., Horvat, Kristijan and Tropea, Cam (2006): *Spray Impact: Rim Transverse Instability Initiating Fingering and Splash, and Description of a Secondary Spray*. Physics of Fluids, 102104, Vol. 18.

Rosskamp, H. (1998): *Simulation von drallbehafteten Zweiphasenströmungen mit schubspannungsgetriebenen Wandfilmen.* PhD Thesis, Institute for Thermal Turbomachinery, University of Karlsruhe.

Samenfink, W. (1997): *Grundlegende Untersuchung zur Tropfeninteraktion mit schubspannungsgetriebenen Wandfilmen.* PhD Thesis, Institute for Thermal Turbomachniery, KIT.

Samenfink, W., Elsaesser, A., Dullenkopf, K. and Wittig, S. (1999): *Droplet Interaction with Shear-Driven Liquid Films: Analysis of Deposition and Secondary Droplet Characteristics.* International Journal of Heat and Fluid Flow, Vol. 20, pp 462 – 469.

Sattelmeyer, T. (1985): *Einfluss der ausgebildeten turbulenten Luft-Fluessigkeitsfilmstroemung auf den Filmzerfall und die Tropfenbildung am Austritt von Spalten geringer Hoehe.* PhD Thesis, Institute for Thermal Turbomachinery, University of Karlsruhe.

Schneider, M. (1989): *Experimentelle Untersuchung zur Tropfen-Filmwechselwirkung.* Studienarbeit, Institute of Thermodynamics, KIT.

Schober, P. (2009): *Berührungsfreie Erfassung beschleunigter schubspannungsgetriebener Kraftstoffwandfilme unter Druckeinfluss.* PhD Thesis, Institute for Thermal Turbomachinery, University of Karlsruhe.

Schotland, R.M. (1960): *Experimental Results Relating to the Coalescence of Water Drops with Water Surfaces.* Discussion. Faraday Soc.Experiments in Fluids, Vol. 30, pp 72–77.

Simmons, K., Hibberd, S., Wang, Y. and Care, I. (2002): *Numerical Study of the Two-Phase Flow within an Aero Engine Bearing Chamber Model using a Coupled Lagrangian Droplet Tracking Method.* Computational Technologies for Fluid systems with Industrial Applications, PVP2002-1568, Vol. 448.

Simmons, K., Johnson, G. and Wiedemann, N. (2011): *Effect of Pressure and Oil Mist on Windage Power Loss of a Shrouded Spiral Bevel Gear.* ASME Conference Proceedings, Vol. 2011, pp 327–335.

Stanton, D. and Rutland, C. (1996): *Modeling Fuel Film Formation and Wall Interaction in Diesel Engines.* SAE Conference Proceedings, Vol. 1996, pp 1249–1257.

Stanton, D. and Rutland, C. (1998): *Multi-Dimensional Modeling of Thin Liquid Films and Spray-Wall Interactions Resulting from Impinging Sprays.* International Journal of Heat and Mass Transfer, Vol. 41, pp 3037 – 3054.

Stow, C. D. and Hadfield, M. G. (1981): *An Experimental Investigation of Fluid Flow Resulting from the Impact of a Water Drop with an Unyielding Dry Surface.* Proceedings of the Royal Society of London. A. Mathematical and Physical Sciences, Vol. 373, pp 419–441.

Stow, C. D. and Stainer, R. D. (1977): *The Physical Products of a Splashing Water Drop.* Meteorological Society of Japan Journal, Vol. 55, pp 518–532.

Stratmann, Jochen (2009): *Droplet-Wall and Spray-Wall Interaction at Increased Ambient Pressure and Wall Temperature*. PhD Thesis, RWTH Aachen.

Taylor, Geoffrey (1959): *The Dynamics of Thin Sheets of Fluid. II. Waves on Fluid SHeets*. Vol. 253, pp 313–321.

Thoroddsen, S.T. (2002): *The Ejecta Sheet generated by the Impact of a Drop*. Journal of Fluid Mechanics, Vol. 451, pp 373–381.

Tropea, C. and Marengo, M. (1999): *The Impact of Drops on Walls and Films*. Multiphase Science and Technology, Vol. 11, pp 19–36.

Trujillo, M. F. and Lee, C. F. (2001): *Modeling Crown Formation due to the Splashing of a Droplet*. Physics of Fluids, Vol. 13, pp 2503–2516.

Ubbink, O. and Issa, R.I. (1999): *A Method for Capturing Sharp Fluid Interfaces on Arbitrary Meshes*. Journal of Computational Physics, Vol. 153, pp 26–50.

van Hinsberg, Nils (2010): *Investigation of Drop and Spray Impingement on a Thin Liquid Layer accounting for the Wall Film Topology*. PhD Thesis, TU Darmstadt.

Vander Wal, R., Berger, G. and Mozes, S. (2006a): *Droplets Splashing upon Films of the Same Fluid of Various Depths*. Experiments in Fluids, Vol. 40, pp 33–52.

Vander Wal, R., Berger, G. and Mozes, S. (2006b): *The Splash/Non-Splash Boundary upon a Dry Surface and Thin Fluid Film*. Experiments in Fluids, Vol. 40, pp 53–59.

Villermaux, E. and Bossa, B. (2011): *Drop Fragmentation on Impact*. Journal of Fluid Mechanics, Vol. 668, pp 412–435.

Walzel, P. (1980): *Zerteilgrenze beim Tropfenprall*. Chemie Ingenieur Technik, Vol. 52, pp 338–339.

Wang, A.-B., Chen, C.-C. and Hwang, W.-C. (2002): *On Some New Aspects of Splashing Impact of Drop-Liquid Surface Interactions*. In: Rein, M. (Publisher), Drop-Surface Interaction, CISM Course and Lectures - No. 456, Springer, Wien/New York.

Weinstock, V.D. and Heister, S.D. (1997): *Review: The Transient Equation of Motion for Particles, Bubbles and Droplets*. Journal of Fluid Engineering, Vol. 119, pp 233–247.

Weinstock, V.D. and Heister, S.D. (2006): *Modelling Oil Flows in Engine Sumps: Drop dynamics and Wall impact Simulation*. Journal of Engineering for Gas Turbines and Power, Vol. 128, pp 163–172.

Weiss, D. and Yarin, A. L. (1999): *Single Drop Impact onto Liquid Films: Neck Distortion, Jetting, Tiny Bubble Entrainment, and Crown Formation*. Journal of Fluid Mechanics, Vol. 385, pp 229–254.

Wieth, L., Braun, S., Koch, R. and Bauer, H.-J. (2014): *Modeling of Liquid-Wall Interaction using the Smoothed Particle Hydrodynamics (SPH)*. ILASS Europe - 2014 <Bremen, Germany,2014>.

Wittig, S. and Busam, S.. (1996): *Lubrication Systems and Seals Technologies*. Techn. rep.

Wittig, S., Glahn, A. and Himmelsbach, J. (1994): *Influence of High Rotational Speeds on Heat Transfer and Oil Film Thickness in Aero-Engine Bearing Chambers*. Journal of Engineering for Gas Turbines and Power, Vol. 116, pp 395–401.

Worthington, A. M. (1876): *On the Forms Assumed by Drops of Liquids Falling Vertically on a Horizontal Plate*. Proceedings of the Royal Society of London, Vol. 25, pp 261–272.

Worthington, A. M. (1879): *On the Spontaneous Segmentation of a Liquid Annulus*. Proceedings of the Royal Society of London, Vol. 30, pp 49–60.

Wurz, D. (1971): *Experimentelle Untersuchung des Stroemungsverhalten duenner Wasserfilme und deren Rueckwirkung auf einen gleichgerichteten Luftstrom maessiger bis hoher Unterschallgeschwindigkeit*. PhD Thesis, Institute for Thermal Turbomachinery, University of Karlsruhe.

Yarin, A.L. (2006): *Drop Impact Dynamics: Splashing, Spreading, Receding, Bouncing*. Annual Review of Fluid Mechanics, Vol. 38, pp 159–192.

Yarin, A. L. and Weiss, D. A. (1995): *Impact of Drops on Solid Surfaces: Self-Similar Capillary Waves, and Splashing as a New Type of Kinematic Discontinuity*. Journal of Fluid Mechanics, Vol. 283, pp 141–173.

Yokoi, K. (2008): *A Numerical Method for Free-Surface Flows and Its Application to Droplet Impact on a Thin Liquid Layer*. Journal of Scientific Computing, Vol. 35, pp 372–396.

Yokoi, K. (2013): *A Practical Numerical Framework for Free Surface Flows based on CLSVOF Method, Multi-Moment Methods and Density-Scaled CSF Model: Numerical Simulations of Droplet Splashing*. Journal of Computational Physics, Vol. 232, pp 252 – 271.

Young, C. and Chew, J. (2005): *Evaluation of the Volume of Fluid Modelling Approach for Simulation of Oil/Air System Flows*. ASME Conference Proceedings, Vol. 2005, pp 1249–1257.

Zhang, L., Brunet, P., Eggers, J. and Deegan, R. (2010): *Wavelength Selection in the Crown Splash*. Physics of Fluids, Vol. 22, pp 105–114.

Zhang, L. V., Toole, J., Fezzaa, K. and Deegan, R. D. (2012): *Evolution of the Ejecta Sheet from the Impact of a Drop with a Deep Pool*. Journal of Fluid Mechanics, Vol. 690, pp 5–15.

Zhang, M.Y., Zhang, H. and Zheng, L.L. (2008): *Simulation of Droplet Spreading, Splashing and Solidification using Smoothed Particle Hydrodynamics Method*. International Journal of Heat and Mass Transfer, Vol. 51, pp 3410 – 3419.

Zhbankova, S.L. and Kolpakov, A.V. (1990): *Collision of Water Drops with a Plane Water Surface*. Fluid Dynamics, Vol. 25, pp 470–473.

Appendix

A.1   First Part

The Dimensional Influence Parameter

General Characteristics

1. Gravitational acceleration g $[m/s^2]$
   The gravitational acceleration with its linear dependence on the potential energy is a parameter that is indeed identified as an influence parameter in literature but is rarely varied in isolation due to the tremendous experimental effort. For the liquid film dynamics a change in gravitational acceleration may have an impact on the cavity and crown dynamics as well as on the spray generation. Kyriopoulos (2010) was one of the few authors found in literature dealing with the effect of gravity. She found a significant influence of gravity on the outcome of spray cooling.

Fluid Properties

The fluid properties of the lubrication oil in bearing chamber are mainly influenced by the internal temperature in oil systems. The internal temperature in oil systems changes depending on the location and the operation point of the aero engine between $253K$ and $523K$, respectively (see in Glahn (1995)). Following the assumption that the impacting drop equals the target surface fluid properties, one set of liquid substance properties are given with:

1. Liquid density ρ_l $[kg/m^3]$
   The linear dependency of the kinetic, potential and dissipation energy on the liquid density enforces its important role in drop impact dynamics. Compressibility effects arising from high pressure become furthermore relevant in the initial stage of the impact evolution. Although the temperature in bearing chambers varies over a wide range during the operation of the aero engine, its influence on the liquid density compared to other fluid appears to be low. Gorse (2007) proposed a correlation stating the dependence of the liquid density ρ_l of a typical lubrication oil on the temperature T $[K]$ with $\rho_l = 1203 - 0.733T$.

2. Dynamic liquid viscosity μ_l $[kg/ms]$
   The liquid viscosity dictates the rate of energy dissipation occurring during the drop impact. The outcome of the impingement process is known to be strongly affected by the liquid viscosity (see in i.e. Macklin and Metaxas (1976), Vander Wal et al. (2006a) and Vander Wal et al. (2006b)). Beside the liquid film dynamics it can also affect the disintegration process (rim instabilities) leading to the spray generation (see in Gueyffier and Zaleski (1998)). In contrast to the liquid density, its dependence on temperature is considerable large. Glahn (1995) proposed for the kinematic viscosity ν_l $[10^{-6}m^2/s]$ a relation to the fluid temperature

with $loglog(\nu_l + 0.8) = 8.962182 - 3.527033logT$. The lubrication fluid viscosity can vary by several orders of magnitude during the operation of an aircraft engine. As can be seen later, this results in a parameter range of the liquid viscosity often outside of the validity of various drop impingement descriptions in the existing literature.

3. surface tension σ $[N/m]$
   Surface tension results in a reset force that aims to put the free surface in its initial state appearing often as an ideal energetic state. This state is reached when a minimum area of the free surface and a nearly constant pressure distribution for the underlying condition occurs. The free surface is strongly deformed in drop impact dynamics thereby assigning the surface tension an important role (Stow and Stainer (1977) and Vander Wal et al. (2006b)). Surface tension is moreover appearing linear dependent on the surface energy. The surface energy dictates mainly the evolution of the free surface after impingement when the fraction of the surface energy gets significant compared to the kinetic and potential energy. Gorse (2007) developed also an empirical correlation for the temperature dependence of the oil surface tension σ $[10^{-3}N/m]$ with $\sigma = 54.665 - 0.077198T$.

Impacting Spray Characteristics

1. Impacting drop diameter D $[m]$
   The impacting drop diameter is one of the most relevant influence parameter in the drop impingement process. Due to the relevance in nearly all the energies determining the drop impact evolution, it has been shown to effect in many investigation the impingement regimes, the liquid film dynamics and the spray generation (Macklin and Metaxas (1976) and Stow and Stainer (1977)). According to the studies of Glahn et al. (1997), Birkenkaemper (1996), Glahn et al. (2002), Ebner et al. (2002) and Glahn et al. (2003), the drop sizes in oil systems are in the submillimeter range.

2. Impacting drop sphericity Ψ $[-]$
   In many theoretical and numerical descriptions the drops are always assumed as perfect spheres. The aerodynamic forces acting on the droplet, oscillations deriving from prior disintegration and the gravitational forces, are often responsible for a slightly deformation of the drop into an ellipsoid. A deformed or oscillating drop may hence also have different local curvatures compared to a perfect sphere at its contact with the liquid surface. This change in local surface energy in turn may have an influence on the liquid film dynamics and spray generation (see in Rodriguez and Mesler (1988)). In bearing chamber, due to the successful application of the PDA measurement technique, nearly spherical droplets occur very frequent. Visualizations from Gorse (2007) proved however also the existence of ligaments and strong deformed drops in bearing chambers.

3. Impacting absolute drop velocity V $[m/s]$
   The impact drop velocity is the most investigated parameter in literature since it can be easily varied with the height of the drop generation source for single drops or with the nozzle pressure for mono-dispersed drop streams and poly-disperse sprays. Numerous

investigations proved the large effect of the impact velocity on the impingement process, i.e. the cavity, crown dynamics and spray generation. Its importance arises from the square dependence on the kinetic energy and the significant role in the dissipated energy during the impact evolution. In oil systems, the detected flow velocities of the droplets can reach up to $30 \; m/s$ depending on operating condition of the aero-engine.

4. Impact angle α [$-$]

   The impingement angle defines the incident direction. It is formulated with the angle between the impacting drop absolute velocity vector and the local wall or liquid surface. For walls having a roughness in the same order of magnitude as the impacting drop diameter, it is required to identify the local inclination of the wall surface. For drop impacts on liquid surfaces the latter information plays a minor role. The regime morphology was investigated by Leneweit et al. (2005) for low impact kinetic energy and by Okawa et al. (2008) for the impingement regimes and the spray generation during single water drop impact. An immense difference between the normal drop impact was found for flatter impact angle while minor differences were identified for steeper angles. Based on the available insights in bearing chamber, it is expected that larger drops are impacting with steeper impact angles onto liquid surfaces. The stronger centrifugal forces of larger drops lead to a sooner leaving of the gas streamlines and hence to enhanced radial velocity components (Farral (2000)). The smaller drops on the other hand follow more the gas streamlines and reach the chamber wall films with much flatter angles (Farral (2000)).

Surrounding Gas Characteristics

1. Gas velocity profile

   The gas velocity driving the liquid film or flowing over the dry wall builds a boundary layer near the free surface. In addition to the aerodynamic forces acting onto the impacting drop, the strong gradient occurring near the wall or the liquid surface may disturb the droplet and lead to a deformation. This in turn can lead to similar effects as they already were mentioned for the sphericity. In Samenfink et al. (1999) droplet interaction with liquid surfaces driven by an air flow of approx. $30 \; m/s$ was investigated. In their publication no clear influence of the gas boundary layer on the impact outcome was identified although existing throughout the investigation. The air flow in oil systems is indeed accelerated by the rotating components leading to complex swirled flows with strong secondary vortices (see e.g. Gorse (2007)).

2. Gas pressure p_g [bar]

   Similar to the gas velocity influence very little research is available for the impact of gas pressure on the impingement evolution. An increase in gas pressure enhances for instance the interaction of the entrapped gas bubbles inside the liquid film due to the increased density (Rein and Delplanque (2008)). On the other hand also slightly differences in the surface tension forces are assumed. For spray impingement onto heated walls, Stratmann (2009) conducted some experiments with variable density. In Engel (1966) on the other hand the gas pressure was reduced in order to diminish the aerodynamic drag and gain higher impact velocities. The gas pressure in oil systems of aero engines can reach values around 6 bar (absolute) depending on the operation of the aircraft engine and the location of the bearing chamber.

Target Surface Characteristics

1. Solid surface

   The impact of droplets onto solid surfaces is mainly influenced by the surface curvature, R_c, the roughness R_a and the wall temperature. The effect of curvature is often measured by the ratio of the impinging drop diameter to the radius of curvature of the surface, D/Rc (see in Levin and Hobbs (1971)). For a small ratio as it is found in oil systems this influence is nearly negligible for drop-film interaction modelling.

   The surface roughness on the other hand has a significant impact on the onset of the splashing regime and the spray generation as found in Stow and Stainer (1977), Walzel (1980), Stow and Hadfield (1981) and Kalantari (2007). As reported in several studies, the wall temperature influence becomes noticeable only near the $Leidenfrost$ temperature. In this regime hydrophobic effects of the surface disappear completely during the impact leading occasionally to pool boiling and droplet evaporation (Richter et al. (2005), Stratmann (2009)). In bearing chamber, it is however expected that drop interaction with solid walls occurs only for a short and unsteady time frame during the film formation. In most of the time, it is expected that the solid surface is covered by a oil liquid diminishing thereby the effect of the solid surface on the impingement evolution.

2. Liquid surface

   A liquid surface exposed to drop impingement changes the dynamics of the impact evolution significantly (Samenfink et al. (1999), Cossali et al. (1997), Tropea and Marengo (1999)). The main influences are the liquid film depth, h_f, the liquid surface structure and the film velocity profile. Through a displacement of the liquid film, kinetic energy is transformed into surface energy and dissipated energy. Hence, less kinetic energy is available for the lamella expansion and breakup of the crown's rim which results in less secondary drops, different size and velocity distribution and ejected mass. Recent investigations demonstrated furthermore that the influence of the liquid film depth is not sufficiently described solely by the relative film height but needs additional information about the impacting energy. Bisighini (2010), for instance, related the maximum cavity depth in deep pool impacts, as a measure of the impingement energy, to the film height, h_f. With this, a new classification for the impingement regimes was proposed. The oil liquid film thickness in bearing chambers can vary between 0.4 mm up to 5 mm depending on where the drop is impacting onto a liquid film.

   The liquid film surface structure could also affect the outcome of a drop impingement in particular when the drop impact angles get flat and the liquid film develops with roll waves thereby evolving with large wave amplitudes. A considerable effect is proved particularly for flat drop impact angles since the majority of the impingements take place on the wave crest rather than on the wave base (see in Samenfink et al. (1999) and van Hinsberg (2010)).

Based on the findings gained from fundamental investigations describing the oil film dynamics in bearing chambers, a strong waviness is expected in the regions outside the vicinity of the scavenge off-take (see in Glahn and Wittig (1996) and Hashmi (2012)). The liquid film velocity and its profile are additionally referred in literature as influence parameter on the impingement outcome though lacking on a quantitative analysis. The majority of the investigations in literature see similarities between drop impingement on a moving and steady liquid film. For this reason an impact onto a moving film is often recreated with the aid of an oblique impact onto a steady liquid film by assimilating the velocity vector summation. This approach is however particularly questionable when the gradient of the liquid film velocity profile near the free surface differs significantly from the constant one in steady impacts, as i.e. in shear driven thin liquid films (Himmelsbach (1992), Samenfink et al. (1999) and Gorse et al. (2004)).

A.2 Second Part

A.3 Spray-Film Interaction Model derived from Single Drop Impingement Data

The spray-impingement model includes the prediction of the threshold between the two regimes, namely deposition and splashing. When a splashing impingement takes place, the model requires to provide the ejected mass fraction and secondary drop characteristics. Hence, the mass and momentum transfer between the liquid film and the impinging drop is characterized. In case of drop deposition, the entire mass and momentum is added to the liquid film.

Classification of the Target Surface

The transition between deposition and splashing depends mainly on the target surface characteristics. Thus, before defining whether a splashing impingement takes place, it is necessary to classify the target surface based on the knowledge derived from the previous chapters. Here, with the help of the maximum cavity depth developing during a deep pool impingement,

$$\Delta_{max,deep} = 0.00225 \cdot We_n^{0.4985} \cdot Oh^{-0.18466} \cdot Fr^{0.2585} \cdot +1.969, \qquad (A.1)$$

the following classification was deduced.

Regime	Range of Validity
Wetted wall	$H^*/\Delta_{max,deep} < 0.1$
Thin films	$0.1 < H^*/\Delta_{max,deep} < 0.4$
Thick films	$0.4 < H^*/\Delta_{max,deep} < 1$

Table A.1: Classification of the target surface for drop impingement onto liquid surfaces

Splashing Threshold

Since the present investigation involved only drop impingement onto liquid surfaces, transition between deposition and splash is determined here only for impacts on liquid surfaces. Now once establishing the character of the target surface, different transition models can be used that have been extensively reviewed in chapter 2. For impingements onto liquid surfaces, the following splashing threshold is derived from literature and own experiments already described in chapter 2.

- $K = We Oh^{-0.4}$,

- $K > 2100$ -> Splashing

- $K < 2100$ -> Spread.

Ejected Mass Fraction

The resulting ejected mass fraction for a splashing impingement is described using two correlations due to the non-linear trend identified with the variation of the impingement angle for $\alpha < 45$. The correlations are summarize in the following equations with

$$M_{sec}/M = \begin{cases} 0.0021092 \cdot We \cdot Oh^{-0.25} \cdot Fr^{0.063} \cdot \left(\frac{H^*}{\Delta_{max,deep}}\right)^{-0.04} \cdot \alpha^{-0.95} + \\ 0.067961, 45 <= \alpha <= 90 \\ 0.0021092 \cdot We \cdot Oh^{-0.25} \cdot Fr^{0.063} \cdot \left(\frac{H^*}{\Delta_{max,deep}}\right)^{-0.04}. \\ (90 - \alpha)^{-0.95} + 0.067961, \alpha < 45. \end{cases} \quad (A.2)$$

Number of Secondary Drops

According to the ejected mass fraction, number of secondary drops expressed alos a non-linear effect with the vaiation of the impingement angle. Therefore, two different definition results also for the number of secondary drops produced from a splashing impingement with

$$N = \begin{cases} 0.018703 \cdot We \cdot Oh^{-0.25} \cdot Fr^{0.063} \cdot \left(\frac{H^*}{\Delta_{max,deep}}\right)^{-0.22} - 27.897, & 45 <= \alpha <= 90 \\ 1.208 \cdot 10^{-5} \cdot We \cdot Oh^{-0.25} \cdot Fr^{0.063} \cdot \left(\frac{H^*}{\Delta_{max,deep}}\right)^{-0.22} \cdot \alpha^2 - 51, & \alpha < 45. \end{cases}$$

$$(A.3)$$

Secondary Drop Sizes

In order to extract the detailed distribution of the secondary spray, two correlations for the arithmetic mean secondary drop size and logarithmic standard deviation are proposed next. The values are then simply introduced into a log-normal distribution function where relative and cumulative number frequency of the secondary spray may be derived. The log-normal distribution function already proved in the experimental and numerical chapter of the present manuscript to fit the measurements with high accuracy.

$$D_{10}/D = 0.3015 \cdot We \cdot Oh^{-0.6} \cdot \alpha^{-3} + 0.069894 \tag{A.4}$$

$$\sigma_{ln,D} = 0.0048535 \cdot We \cdot Oh^{-0.4} \cdot Fr^{0.05} \cdot \alpha^{-1.25} + 0.15585 \tag{A.5}$$

Secondary Drop Velocities

The parallel and normal component of the secondary drop velocity is also provided by the present spray impingement model. Note, that for non-normal impingements all secondary drops are released only in downstream direction of the impingement as characterized in the previous section.

$$V_{mean,pll,sec}/V = 1788 \cdot We^{-1.0292} \cdot Oh^{0.288} \cdot (\frac{H^*}{\Delta_{max,deep}})^{0.10292} \tag{A.6}$$

$$V_{mean,orth,sec}/V = 0.00637 \cdot We \cdot Oh^{-0.4} \cdot \alpha^{-1} + 0.45068 \tag{A.7}$$

Position of Ejection

A simplified way to find the time and position of the produced secondary drops can be derived from the maximum lamella height, the time to reach the max. lamella height and the lamella diameter resulting at max. crown height in the following. These correlation proved very accurate for normal drop impingement. The former showed to be of more complex character in non-normal impingements. It may however be adopted as a first approximation.

$$H_{cr,max}/D = 0.000372 \cdot We_n \cdot Oh^{-0.19} \cdot Fr^{0.05} \cdot (\frac{H^*}{\Delta_{max,deep}})^{-0.1}, \tag{A.8}$$

$$\tau_{cr,max}/D = 0.00712 \cdot We_n \cdot Oh^{-0.1} \cdot Fr^{0.04} \cdot (\frac{H^*}{\Delta_{max,deep}})^{-0.13} - 2.9406, \tag{A.9}$$

$$D_{cr,max}/D = 0.95 \cdot (\tau_{cr,max} - \tau_0)^{0.34835 \cdot (\frac{H^*}{\Delta_{max,deep}} - 0.13543)^{-0.032585}}, \tag{A.10}$$

Range of Validity of the Correlations

The described correlations are tested and valid for the following range of influence parameter.

- $300 < We < 2376$
- $0.0018 < Oh < 0.047$
- $136 < Fr < 282928$
- $0.1 < H^*/\Delta_{max,deep} < 1$
- $30° < \alpha < 90°$

Lebenslauf und persönliche Daten

Name	Davide Peduto
Geburtsdatum	20. Mai 1983
Geburtsort	Heidelberg
Familienstand	ledig
1989-1994	Gerhart Hauptmann Grundschule, Mannheim
1994-2003	Kurpfalz Gymnasium, Mannheim
2003	Allgemeine Hochschulreife
2003-2009	Studium des Maschinenbaus an der Universität Karlsruhe (TH)
2005-2006	Werkstudent bei Voith Industrial Services Engineering GmbH, Ludwigshafen
2006	Vordiplom
2006-2007	Hilfsassistent am Institut für Thermische Strömungsmaschinen der Universität Karlsruhe (TH)
2007	Praktikum bei Voith Siemens Hydro Power Generation, Heidenheim
2008-2009	Diplomarbeit am Rolls-Royce University Technology Centre, Nottingham, GB
2009	Diplom
2009-2013	Wissenschaftlicher Mitarbeiter am Institut für Thermische Strömungsmaschinen des Karlsruher Instituts für Technologie (KIT)
2011-2014	Begleitstudium Innovation, Business Creation und Entrepreneurship an der KIC InnoEnergy PhD School, Europa
2009-2015	Teilzeit Doktorand im Rahmen einer Kooperationspromotion am Rolls-Royce University Technology Centre der Universität Nottingham, GB
Seit 01.10.2013	Entwicklungsingenieur bei BorgWarner Turbo Systems Engineering GmbH, Kibo im Bereich Core Science Bearings & Acoustics